RECOMBINANT DNA SAFETY CONSIDERATIONS

Safety considerations
for industrial, agricultural and environmental
applications of organisms
derived by recombinant DNA techniques

ORGANISATION FOR ECONOMIC CO-OPERATION AND DEVELOPMENT

Pursuant to article 1 of the Convention signed in Paris on 14th December, 1960, and which came into force on 30th September, 1961, the Organisation for Economic Co-operation and Development (OECD) shall promote policies designed:

- to achieve the highest sustainable economic growth and employment and a rising standard of living in Member countries, while maintaining financial stability, and thus to contribute to the development of the world economy;
- to contribute to sound economic expansion in Member as well as non-member countries in the process of economic development; and
- to contribute to the expansion of world trade on a multilateral, non-discriminatory basis in accordance with international obligations.

The Signatories of the Convention on the OECD are Austria, Belgium, Canada, Denmark, France, the Federal Republic of Germany, Greece, Iceland, Ireland, Italy, Luxembourg, the Netherlands, Norway, Portugal, Spain, Sweden, Switzerland, Turkey, the United Kingdom and the United States. The following countries acceded subsequently to this Convention (the dates are those on which the instruments of accession were deposited): Japan (28th April, 1964), Finland (28th January, 1969), Australia (7th June, 1971) and New Zealand (29th May, 1973).

The Socialist Federal Republic of Yugoslavia takes part in certain work of the OECD (agreement of 28th October, 1961).

Publié en français sous le titre:

CONSIDÉRATIONS DE SÉCURITÉ
RELATIVES À L'ADN RECOMBINÉ

This study was undertaken as a follow-up to the OECD report *Biotechnology: International Trends and Perspectives*, in accordance with the wish expressed by the Committee for Scientific and Technological Policy during its 34th Session of 10th-11th February, 1983. It was carried out by an *Ad hoc* Group of government experts created by that Committee in July 1983, with the assistance of the OECD Secretariat.

The OECD Council decided on 30th May, 1986 to make the report of the *Ad hoc* Group publicly available.

On 16th July, 1986 the Council adopted the Recommendation which appears on the following pages and which is based on Chapter V of the report.

OECD Council Recommendations are acts made by mutual agreement of all the Members which are submitted to them for consideration in order that they may, if they consider it opportune, provide for their implementation.

Also available

BIOTECHNOLOGY AND PATENT PROTECTION. An International Review by
F.K. Beier, R.S. Crespi and J. Straus (September 1985)
(93 85 05 1) ISBN 92-64-12757-7   134 pages   £8.00   US$16.00   F80.00   DM35.00

SCIENCE AND TECHNOLOGY POLICY OUTLOOK – 1985 (June 1985)
(92 85 03 1) ISBN 92-64-12738-0   98 pages   £5.50   US$11.00   F55.00   DM24.00

BIOTECHNOLOGY. International Trends and Perspectives by Alan T. Bull,
Geoffrey Holt, Malcolm D. Lilly (September 1982)
(93 82 01 1) ISBN 92-64-12362-8   84 pages   £5.50   US$11.00   F55.00   DM28.00

Prices charged at the OECD Bookshop.

*THE OECD CATALOGUE OF PUBLICATIONS and supplements will be sent free of charge
on request addressed either to OECD Publications Service, Sales and Distribution Division,
2, rue André-Pascal, 75775 PARIS CEDEX 16, or to the OECD Sales Agent in your country.*

TABLE OF CONTENTS

5

RECOMMENDATION OF THE COUNCIL

CONCERNING SAFETY CONSIDERATIONS FOR APPLICATIONS OF RECOMBINANT DNA* ORGANISMS IN INDUSTRY, AGRICULTURE AND THE ENVIRONMENT

The Council,

Having regard to Articles 1 *(c)*, 3 *(a)* and 5 *(b)* of the Convention on the Organisation for Economic Co-operation and Development of 14th December 1960;

Having regard to the report *Recombinant DNA Safety Considerations – Safety Considerations for Industrial, Agricultural and Environmental Applications of Organisms derived by Recombinant DNA Techniques*;

Considering that recombinant DNA techniques have opened up new and promising possibilities in a wide range of applications and can be expected to bring considerable benefits to mankind;

Recognising, in particular, the contribution of these techniques to improvement of human health and that the extent of this contribution is expected to increase significantly in the near future;

Considering that a common understanding of the safety issues raised by recombinant DNA techniques will provide the basis for taking initial steps toward international consensus, the protection of health and the environment, the promotion of international commerce and the reduction of national barriers to trade in the field of biotechnology;

Considering that the vast majority of industrial recombinant DNA large-scale applications will use organisms of intrinsically low risk which warrant only minimal containment consistent with good industrial large-scale practice (GILSP);

Considering that the technology of physical containment is well known to industry and has successfully been used to contain pathogenic organisms for many years;

Recognising that, when it is necessary to use recombinant DNA organisms of higher risk, additional criteria for risk assessment can be identified and that these organisms can also be handled safely under appropriate physical and/or biological containment;

7

Considering that assessment of potential risks of recombinant DNA organisms for environmental or agricultural applications is less developed than the assessment of potential risks for industrial applications;

Recognising that assessment of potential risk to the environment of environmental and agricultural applications of recombinant DNA organisms should be approached with reference to, and in accordance with, information held in the existing data base, gained from the extensive use of traditionally modified organisms in agriculture and the environment generally, and that with step-by-step assessment during the research and development process potential risk should be minimised;

Considering the present state of scientific knowledge;

Recognising that the development of general international guidelines governing agricultural and environmental applications of recombinant DNA organisms is considered premature at this time;

Recognising that there is no scientific basis for specific legislation to regulate the use of recombinant DNA organisms;

On the proposal of the Committee for Scientific and Technological Policy:

1. RECOMMENDS that Member countries,

    a) share, as freely as possible, information on principles or guidelines for national regulations, on developments in risk analysis and on practical experience in risk management with a view to facilitating harmonization of approaches to recombinant DNA techniques;

    b) examine their existing oversight and review mechanisms to ensure that adequate review and control of the implementation of recombinant DNA techniques and applications can be achieved while avoiding any undue burdens that may hamper technological developments in this field;

    c) recognise, when aiming at international harmonization, that any approach to implementing guidelines should not impede future developments in recombinant DNA techniques;

    d) examine at both national and international levels further developments such as testing methods, equipment design, and knowledge of microbial taxonomy to facilitate data exchange and minimise trade barriers between countries. Due account should be taken of ongoing work on standards within international organisations, e.g. WHO, CEC, ISO, FAO, MSDN[1];

    e) make special efforts to improve public understanding of the various aspects of recombinant DNA techniques;

    f) watch the development of recombinant DNA techniques for applications in industry, agriculture and the environment, while recognising that for certain industrial applications, and for environmental and agricultural applications of recombinant DNA organisms, some countries may wish to have a notification scheme;

g) ensure that assessment and review procedures protect intellectual property and confidentiality interests in applications of recombinant DNA, recognising the need for innovation while still ensuring that all necessary information is made available to assess safety.

2. RECOMMENDS, *with specific reference to industrial applications*, that Member countries:

a) ensure, in large-scale industrial applications of recombinant DNA techniques, that organisms which are of intrinsically low risk are used wherever possible, and handled under the conditions of Good Industrial Large-Scale Practice (GILSP) described in the report;

b) ensure that, when a risk assessment using the criteria defined in the report indicates that a recombinant DNA organism cannot be handled merely by GILSP, appropriate containment measures, in addition to GILSP, and corresponding to the risk assessment are applied.

c) encourage, in large-scale industrial applications requiring physical containment, further research to improve techniques for monitoring and controlling non-intentional release of recombinant DNA organisms.

3. RECOMMENDS, *with specific reference to agricultural and environmental applications,* that Member countries:

a) use the existing considerable data on the environmental and human health effects of living organisms to guide risk assessments;

b) ensure that recombinant DNA organisms are evaluated for potential risk, prior to applications in agriculture and the environment by means of an independent review of potential risks on a case-by-case basis[2];

c) conduct the development of recombinant DNA organisms for agricultural or environmental applications in a stepwise fashion, moving, where appropriate, from the laboratory to the growth chamber and greenhouse, to limited field testing and finally, to large-scale field testing;

d) encourage further research to improve the prediction, evaluation, and monitoring of the outcome of applications of recombinant DNA organisms.

4. INSTRUCTS the Committee for Scientific and Technological Policy to:

a) review the experience of Member countries in implementing the principles contained in the report;

b) review actions taken by Member countries in pursuance of this Recommendation and to report thereon to the Council;

c) consult with other appropriate Committees of the OECD in developing proposals for a co-ordinated future work programme in biotechnology.

NOTES AND REFERENCES

*   Deoxyribonucleic acid (see Glossary).

1. World Health Organisation (WHO); Commission of the European Communities (CEC); International Standards Organisation (ISO); Food and Agriculture Organisation (FAO); Microbial Strains Data Network (MSDN).

2. Case-by-case means an individual review of a proposal against assessment criteria which are relevant to the particular proposal; this is not intended to imply that every case will require review by a national or other authority since various classes of proposals may be excluded.

TRANSMITTAL LETTER

to the Chairman of the Committee
for Scientific and Technological Policy

The mandate set by the Committee for Scientific and Technological Policy in July 1983 presented the *Ad hoc* Group on Safety and Regulations in Biotechnology with a considerable challenge when it began its work in December of that year.

It has been an extremely rewarding experience for me to chair such a large and distinguished group of scientists, industrialists, policy makers and representatives of regulatory agencies in their deliberations on an issue central to the continued safe development of the "new" biotechnology – in particular the practical applications of recombinant DNA (rDNA) organisms. By way of a personal review, I would like to make a few observations on our work.

You will be aware that our study has benefited from robust and vigorous discussion and debate and that this finally led to unanimous agreement on our report which provides a framework for the assessment and control of any risks that may be associated with rDNA applications. I, therefore, submit our report to you, confident in the knowledge that the issues have been thoroughly aired and real agreement reached between experts.

We know that the OECD countries and a number of international Organisations are eagerly awaiting the result of our work. We know also that field experiments with rDNA organisms for release into the environment are planned for the near future. In this context, there was considerable pressure to finish our report, more so as many countries were moving to finalise their regulatory policies for biotechnology safety. Your Committee's initiative could not have been more timely.

I would like to stress a number of important points about our report:

- *First,* we have paid special attention to the central issue of our mandate, that of identifying scientific criteria for the safe use of rDNA organisms in industry agriculture and the environment. We have conducted only preliminary analysis of other issues raised by our mandate, that of reviewing the existing regulatory schemes and approaches of OECD countries to biotechnology safety generally. In our view additional resources would be required to complete satisfactorily our initial survey of Member country positions;

- *Second,* in our report we have devised a general scientific framework for risk assessment of rDNA applications in industry, agriculture and the environment. The more detailed scientific and technical considerations presented in the Appendices should also be a useful tool for biotechnology practitioners in establishing their own procedures for risk assessment and risk management;

11

- **_Third,_** we recognise that for environmental and agricultural applications in particular the final establishment of internationally agreed safety criteria may be some way off and that further research may be needed. Accordingly, we suggest a provisional approach incorporating independent case by case review of the potential risks of such proposals prior to application;

- **_Fourth,_** acknowledging that the vast majority of industrial rDNA applications to date have used organisms of intrinsically low risk, we propose a corresponding level of control, "Good Industrial Large Scale Practice (GILSP)" based on existing good industrial practices. Criteria for these organisms of low risk are identified. For rDNA organisms of higher risk, we have also described additional control and containment options.

The Group strongly believes that the acceptance of this internationally agreed framework is an important step. We hope it will facilitate the development of biotechnology applications, with their increasing promise of considerable benefit to mankind, and that at the same time it will help to ensure that due regard is paid to any potential concern.

Finally, I would like to acknowledge both the commitment of all members of the Group to the task of producing a scientifically accurate and balanced report and the support of the OECD Secretariat. It was a pleasure to work with such an exceptional Group.

We are, Sir, at your disposal to assist you in any follow-up if you deem this necessary.

Yours sincerely,

Roger Nourish
Chairman of the _Ad hoc_ Group on
Safety & Regulations in Biotechnology

INTRODUCTION

The focus of this report is on the industrial, agricultural and environmental applications of recombinant DNA engineered organisms[1] because rDNA techniques have developed and progressed to the stage of commercialisation during the last fifteen years. With the emerging commercial use of rDNA organisms, questions arose regarding the capacity of present approaches to address incremental risks associated with these new techniques. The public debate which has ensued has sometimes been characterised by a lack of sufficient understanding of recent scientific advances.

The work of the *Ad hoc* Group centred on industrial, agricultural and environmental applications and attempted to provide some answers to these questions. The main tasks included in the mandate of the *Ad hoc* Group[2] were to:

"*i)* Review country positions as to the safety in use of genetically engineered organisms at the industrial, agricultural and environmental levels, against the background of existing or planned legislation and regulations for the handling of micro-organisms;

ii) Identify what criteria have been or may be adopted for the monitoring or authorisation for production and use of genetically engineered organisms in:

- Industry;
- Agriculture;
- The environment;

Explore possible ways and means for monitoring future production and use of rDNA organisms in:

- Industry;
- Agriculture;
- The environment".

A common understanding of the safety issues raised by rDNA techniques will provide the basis for taking initial steps toward international accord, the protection of health and of the environment, the promotion of international commerce, and the reduction of national barriers to trade in the field of biotechnology.

Biotechnology is not new. Before 3000 BC the Sumerians exploited the ability of yeast to make alcohol in the form of beer. Pharmaceutical, fermentation, and agricultural industries were developed years ago, in part by successfully utilising biotechnology on a commercial scale, which they continue to do today. These traditional industrial and agricultural applications of biotechnology are currently regulated in many countries. Industry and agriculture have safely and successfully used conventional methods of genetics on a commercial scale for decades (e.g. natural selection, cross breeding, conjugation, chemical or radiation induced mutation, transformation).

In the past fifteen years a new dimension was added to biotechnology when scientists discovered biological techniques to recombine *in vitro* the DNA from different organisms. The first, and best known technique, is recombinant DNA (rDNA). It has been the subject of intense research and development during the past ten years and has been shown to be safe when used in the laboratory. The first commercial applications have been approved (e.g. human insulin, phenylalanine, human growth hormone).

Recombinant DNA techniques represent a development of conventional procedures. They permit precise alteration, construction, recombination, deletion and translocation of genes that may give the recipient cells a desirable phenotype. Moreover, rDNA techniques allow genetic material to be transferred into, and to express in, another organism which may be quite unrelated to the source of the transferred DNA.

The following are a number of the concepts the *Ad hoc* Group emphasized in developing criteria for considering rDNA organisms and their applications:

- The *Ad hoc* Group chose to focus this document only on the issues relating to those products and processes developed by the most well-described techniques of rDNA. However, rDNA considerations in this report may also apply to organisms modified by other techniques of genetic manipulation. These would include conventional techniques of genetic manipulation (crossing, conjugation, mutagenesis, selection etc.) and *in vitro* techniques such as cell and protoplast fusion, embryo transfer, and micro-injection;
- The *Ad hoc* Group also decided that it would not deal with genetic manipulation techniques with direct application to man. The focus of this report is on safety and does not address ethical questions which are a totally different issue;
- The *Ad hoc* Group concentrated on the second task of the mandate, i.e. the identification of criteria and ways and means for monitoring future use of rDNA organisms.

As to the first task of the mandate, we instituted a survey of country positions by means of a questionnaire. In most OECD Member countries there is a broad array of existing legislation relating to health, safety and environmental protection, which could be applied in principle to manage the risks which might arise from the industrial, environmental and agricultural applications of rDNA techniques. General legal provisions affecting the safety of biotechnological processes and products already exist in wide variety. In addition, specific provisions for the application of rDNA techniques can be found in the form of voluntary guidelines or recommendations. Most OECD Member countries have begun to examine these existing oversight mechanisms to ensure that adequate review and control can be applied and to avoid any undue burdens that may hamper technological developments in this field.

This report has been developed to present scientific principles that could underlie risk management approaches in dealing with rDNA techniques in industrial, agricultural and environmental applications. The range of scientific considerations discussed can be used to develop safety policies for rDNA processes. They are not standards for regulating applications or products of rDNA, but a first step towards recommendations for eventual international harmonization on these issues.

Industries that use biotechnology processes in OECD countries have maintained safety through adherence to good industrial practices that favour the use of low risk micro-organisms. Additional appropriate controls and containment boundaries are used to ensure the safe application of pathogenic micro-organisms. Similar approaches are appropriate to industrial uses of micro-organisms achieved through rDNA techniques. It is possible, however, that a situation may arise when there are no appropriate safety criteria. Until firm criteria can be developed, the safety evaluation of a particular application of a rDNA derived organism can be conducted on a case-by-case basis[3].

Different issues may arise when rDNA micro-organisms are proposed for application in the environment. Assessment of potential risks of micro-organisms for environmental or agricultural applications is less developed than the assessment of potential risks for industrial applications. Additional research may be necessary to increase our ability to predict the outcome of introductions of rDNA micro-organisms into ecosystems. Therefore, it is not yet possible to develop data requirements and assessment criteria, internationally. Rather, this report presumes a provisional approach for identifying the information which is required for evaluating potential risks. The group felt that this provisional approach would confer sufficient flexibility to suit individual countries. As our knowledge increases we expect to see the establishment of internationally agreed safety criteria.

NOTES AND REFERENCES

1. Subsequently referred to in this document as rDNA organisms. For definitions of recombinant DNA see Appendix A.
2. The full text of the mandate is given in Appendix H.
3. See note 2 to the Recommendation of the Council.

Chapter I

THE APPLICATIONS OF RECOMBINANT DNA TECHNIQUES

Large-scale Industrial Applications

This section gives an overview of the current and potential applications of rDNA techniques in industry. We have considered the type and range of opportunities which rDNA techniques can offer in industry, but we have not attempted to provide a comprehensive review of present and foreseeable industrial applications. Useful reviews have been conducted by the U.S. Congress Office of Technology Assessment[1], the World Health Organisation[2], the German Society of Chemical Technology, Chemical Engineering and Biotechnology (DECHEMA)[3] and the OECD[4].

The impact of the application of rDNA techniques has been to open up new possibilities in a wide range of industries. There are opportunities for utilising these techniques in human and animal therapeutics and nutrition, in industrial fermentation processes, in degradation of environmental pollutants, in mineral and oil extraction and in agriculture. Such applications are providing significant new developments in our understanding and diagnosis of disease, which will lead inevitably to better therapeutic and protective methods in health care, and will result in improved industrial and environmental control processes and increased production of higher quality food and fibre.

Manufacturing processes for pharmaceuticals, food additives, fine chemicals and certain bulk chemicals are benefiting in both yield and purity from new biosynthetic processes which are being applied to both old and new products.

Tangible benefits from recombinant DNA techniques are already emerging in the manufacturing industry associated with health care. For example, the first marketed pharmaceutical rDNA product, human insulin, has made available a potentially unlimited supply of human insulin to replace, where appropriate, non-human insulins produced from pig and cow tissues.

The production of insulin demonstrates the capability of rDNA techniques to produce new and perhaps better versions of existing drugs. Of possibly greater significance is the use of these techniques for the production of hitherto scarce and expensive therapeutic substances, such as interferons, blood factors and vaccines, which could not be produced by previously available technology in the necessary quantity and purity. Human growth hormone (hGH), for example, which is used to increase the growth of patients who would otherwise suffer from pituitary dwarfism, was previously available only from human pituitary glands collected after death and insufficient to meet the demand. Using rDNA techniques, the DNA sequence coding for the human growth hormone has now been inserted into bacteria enabling the production of virtually unlimited supplies, reducing the cost. Moreover, the possibility of contamination with harmful adventitious agents (such as the recent finding of Creutzfeldt-Jakob agent in certain pituitary-derived preparations of hGH) has been virtually eliminated.

Many recombinant organisms engineered to produce other scarce pharmaceuticals have been grown on a pilot scale and some on a manufacturing scale, resulting in the production of volumes up to several thousands of litres of bacterial culture. These processes are producing sufficient material to undertake clinical trials in several therapy areas. Scale up of these processes from the laboratory level, through the pilot scale and finally up to manufacturing levels, has been achieved by making use of very extensive experience within industry of large scale fermentation of bacteria, yeasts, and fungi.

The importance of this huge resource of existing experience in fermentation techniques must not be underestimated in the effective harnessing of the new recombinant DNA techniques to produce sufficient quantities of useful products. Moreover, organisms produced by rDNA techniques are likely to be better defined, thereby resulting in the production of highly purified products of very high quality. The combination of existing fermentation with recombinant DNA techniques offers a powerful means of manufacturing many important substances for use in health care and elsewhere.

Work will continue on the development of a range of recombinant DNA products, including a variety of human hormones, modulators of the immune system, anti-viral and anti-cancer agents. Such substances will give us new insight and new therapies for many important diseases. Disease may be diagnosed more accurately and earlier than at present as new diagnostic methods based on rDNA become available. New agents will be produced which prevent or control diseases involving blood clotting mechanisms. Work on producing human "fibrinolytic" enzymes such as urokinase and tissue plasminogen activators and on the large scale expression of the blood coagulation factors IX and VIIIC by rDNA techniques will continue and expand. The latter process will not only provide an increased supply of the scarce coagulation factors but can also assure the purity of the product and reduce the possibility of contamination with adventitious agents. The techniques also provide an opportunity to improve existing vaccines, e.g. for influenza and cholera and to produce entirely new vaccines, e.g. for malaria, herpes and viral haemorrhagic fevers. In all of these cases the need for handling pathogens in large quantities would be eliminated.

Thus, the health care area will be provided with new versions of old drugs, completely new drugs, and vaccines for diseases which at present have poor or non-existent remedies. These developments will be supported by better diagnostic procedures and should lead to better health care for the community as well as for the individual. Economic benefits may be slow to appear but will increase as technology advances.

Within as well as outside the medical area, rDNA techniques are most likely to be utilised in industry towards the yield improvement of known products or the production of novel compounds in the pharmaceutical, chemical, and food industries. Examples include: (a) improvements in the energetics of single cell protein (SCP) production which would lower costs; (b) increased yields of penicillin acylase for the conversion of penicillins to 6-amino penicillanic acid; (c) improved production of the vitamin riboflavin; (d) the generation of novel micro-organisms for the degradation and bioconversion of chlorinated aromatic compounds for treatment of toxic wastes; and (e) increased methane production from refuse and sewage.

The production of enzymes by rDNA techniques for large-scale use in fermenters is having an impact on the manufacturing industry, as are biosynthetic methods for the production of amino acids. Through the power of the recombinant DNA techniques many new developments can be expected to improve the yield and also the quality of existing

17

manufacturing processes. Previously, such applications were accomplished by extensive chance-based classical mutation programmes, which, although successful in many instances, are essentially indiscriminate and random, necessitating expensive screening programmes to select the most suitable organisms.

The new developments have resulted from advances in basic scientific research which allows us to extract or synthesize the DNA sequences for the production of modified organisms and substances. The new rDNA techniques will spread throughout a wide range of manufacturing industries, offering considerable benefits. From the experience already gained, it is clear that the exploitation of rDNA techniques depends heavily on conventional processes such as large scale fermentation or product extraction and purification which, when combined with the means available for product characterisation and quality control, will lead to increased yield, improved purity and greater safety of such products.

Agricultural and Environmental Applications

This section gives an overview of current and potential applications of rDNA techniques in agriculture and the environment. Many of these applications are in the early stages of development; however, they suggest the broad range of opportunities which may be realised in the future.

1. *Recombinant DNA techniques applied to agriculture*

Recombinant DNA techniques are certain to have major applications in agriculture. Many university, government, and corporate research groups around the world are attempting to apply these new techniques with the overall objective of increasing the quantity, quality, and efficiency of food production. Agricultural applications to create new strains of crop plants, new plant and animal diagnostic tools, animal modification, animal vaccines, new pesticides and herbicides are under intensive development.

Classical genetic techniques of selection, isolation, and cross-breeding have dominated agriculture since the transition from hunting and foraging to the cultivation of crops about 10 000 years ago. These approaches remain the backbone of agriculture, enhanced by farm management practices, economies of scale, *in vitro* culturing techniques, and use of chemical pesticides, artificial fertilizers, and extensive irrigation.

Some 29 basic crop species (8 cereals, 3 root crops, 2 sugar crops, 7 grain legumes, 7 oil seeds, plus bananas and coconuts), supplemented by about 15 major vegetable species and 15 fruit crop species, account for the plant products that provide 93 per cent of the human diet. Recombinant DNA techniques now permit entirely new kinds of basic studies of the physiology, biochemistry, and development of these various plant crops and their interaction with such complex agroecosystem factors as beneficial and parasitic micro-organisms and animals, salinity, acidity, moisture, temperature, chemical fertilizers and pesticides.

The specific aims of rDNA techniques in agriculture are, for example, to reduce vulnerability to environmental stresses; to detect and control infectious agents in animals and in the field and post-harvest; to reduce dependence on and modify use patterns of chemical pesticides; to decrease dependence on chemical fertilizers and irrigation; and to increase the nutritional qualities of seeds, fruits, grains, and vegetables.

Recombinant DNA techniques are feasible for manipulations of the micro-organisms important in agriculture and are likely to be well-developed soon for simply-inherited traits in plants and animals. The manipulation of complex inherited traits such as yield, habitat

18

adaptation and photosynthesis, which require expression of many genes, is difficult and not technically feasible at present. Increasing genetic diversity by introducing genes from other sources should improve the ability to compensate for the effects of major environmental changes. The following applications illustrate the directions of this work.

a) *Enhancing the nutritional quality of seed storage proteins*

Seeds of legumes and cereal grains provide approximately 70 per cent of the human dietary protein requirement. However, proteins from these sources do not provide a balanced diet because they are deficient in certain amino acids that humans must obtain from the diet. In the U.S. the first request for field tests of rDNA containing plants in agriculture involved experiments to construct a storage protein gene in *Zea mays* (corn) that would code for protein with adequate dietary amounts of the deficient amino acids.

b) *Increasing resistance to cold temperature and frost damage*

Frost damage accounts for losses estimated at up to $3 billion per year in the United States. Physical methods commonly used to protect valuable crops include wind machines, smudge pots burning fossil fuels, and pumping in large amounts of water. These are all costly, often environmentally undesirable, and rather ineffective. At temperatures within 5°C of the freezing temperature, formation of ice on leaves is mediated by surface proteins of bacteria such as *Pseudomonas syringae* or *Erwinia herbicola*, a process called ice-nucleation[5]. Bacteria deficient in the ice-nucleation process (I-bacteria) may be used to control frost damage in crops by competing with I+ bacteria on leaf surfaces. Such I- bacteria have been *(i)* isolated from natural populations, *(ii)* induced by chemical mutagenesis, and *(iii)* produced by deletion of the ice-nucleation gene by rDNA techniques. The rDNA organism is likely to be at least as well defined as the other mutants and to have a higher probability of competing with strains adapted for a particular host plant.

c) *Enhancing the resistance of crops to chemicals and disease*

Plant growth-controlling chemicals such as herbicides are widely used in agriculture to eliminate weeds that compete with crop plants. Certain important crop plants, such as corn (maize), are resistant to the herbicide atrazine, whereas other crop plants grown in the same agroecosystem, such as soybeans, are damaged by this herbicide. Detoxification of atrazine in corn and resistant weeds is under genetic control. If the detoxification genes could be isolated from resistant weeds and transferred into soybeans by rDNA techniques and if the genes would be expressed in soybeans, the resistance mechanism could also be transferred.

Genes that confer resistance to disease agents and insect pests have often been found in species related to cultivated crop plants. Through interspecific hybridization by conventional crossing techniques, resistance genes have been introduced into crop plants, but not without the simultaneous introduction of many undesirable genes which can only be eliminated by several generations of backcrossing and selection. In cases in which biological barriers prevent cross-fertilization or development of the zygotes, cell and protoplast fusion, followed by regeneration into plants, has enabled transfer of genes across these barriers. Such organism and cell breeding techniques, in combination with rDNA techniques for direct introduction of specific resistance genes into cells or protoplasts, may save many years of conventional plant breeding.

d) *Replacing chemical pesticides with microbial agents*

The micro-organism *Bacillus thuringiensis* has been widely applied as a microbial control agent for several lepidopteran insect pests, such as the gypsy moth. The toxic agent is a protein, the gene for which has been cloned and characterised, which is lethal to larval forms of the insects. Using rDNA techniques, the gene for the toxin has been engineered into bacteria whose habitat is roots in soil, in an attempt to deliver the toxin to a site where one pest is likely to feed. This approach of biological control agents has considerable promise for alternatives to chemical pesticides.

Finally, microbial products also can modify pH or other properties of silage or hay so as to dramatically reduce post-harvest losses due to contaminating organisms. Recombinant DNA techniques may be used to increase the efficiency of these microbial products.

e) *Saving fertilizer through biological nitrogen fixation*

Nitrogen, an essential plant nutrient and a key determinant of crop productivity, is rapidly depleted from soils. Some 60 million metric tons of nitrogen fertilizer are applied annually world-wide, which is projected to increase to 160 million metric tons by the year 2000. Ironically, plants are bathed in nitrogen (80 per cent of air). Unlike nearly all other economic plants, soybeans, alfalfa and other legumes have evolved a symbiotic relationship with *Rhizobium* bacteria in their root nodules to extract nitrogen directly from the air. Assays for nitrogen fixation using reduction of acetylene have made it convenient to screen organisms for nitrogen-fixing activity and to isolate the enzymes involved.

With rDNA techniques, at least 15 different genes have been identified which specify the enzymes, electron transfer proteins, and other proteins involved in nitrogen fixation or regulating their expression. It is feasible to isolate gene clusters, insert them in vectors, and transfer the genes to non-nitrogen-fixing species or genera. Unfortunately, it has been much harder to determine how to make this number of genes function in the recipient cells. It is very likely that significant progress will be made introducing nitrogen fixation into other symbiotic bacteria.

One promising approach employs *Azotobacter*, which is free-living in the soil and does not need to form nodules in the roots of the crop plant. Other soil microbes may be engineered to fix nitrogen.

f) *Diagnosis of plant infections*

Recombinant DNA techniques may be used in diagnosing plant infections. An application of this rDNA technique has permitted useful diagnosis of several virus and viroid-induced diseases. Sequences specific for RNA or DNA of the causal agent have been isolated and diagnostic gene probes prepared. There are also a number of diagnostic methods based on monoclonal antibodies for bacterial (e.g., crown gall disease on greenhouse-grown ornamentals; *Xanthomonas* diseases of citrus and sorghum) and fungal (e.g., *Fusarium* on turf grass) as well as viral disease agents. This approach, which should be very useful in selecting pathogen-free plants for culturing and crosses and in screening crops for infection, has its counterparts in animal and human medicine.

g) *Animal productivity*

In the OECD countries, in particular, a very significant portion of the diet is contributed by meats, fish and poultry. These are very competitive industries, where efficiency of production is highly valued. Animal growth hormone genes have been cloned, put into expression vector systems, and introduced into embryos, leading to faster growth and larger animals. In general, gene transfer in food animals is expected to be undertaken with the intent of introducing commercially valuable characteristics such as increased disease resistance, more rapid growth, qualitative changes in the host animal tissues, etc. Gene transfer in laboratory animals is currently accomplished by mechanical means such as microinjection into embryonic cell masses. In the future, it is possible that gene transfer in food animals will also be accomplished by the use of animal viruses that are sufficiently infective to transfer genetic material but sufficiently temperate so as not to cause overt disease in the host animal.

Diagnosis, treatment, and prevention of animal diseases also have high priority. The world-wide vaccine market for veterinary biologicals is estimated to be $541 million with $118 million going to rabies vaccines, $118 million for poultry products, $109 million for cattle, and the remainder more or less evenly distributed for vaccines for swine, sheep, horses, and cats. Already in use are vaccine products from rDNA techniques to prevent diarrhea in newborn swine (scours). Vaccines against other organisms are also under development including foot-and-mouth disease virus, *Vibrio* species pathogenic for salmon and eels, avian coccidiosis vaccine, and feline leukaemia virus. The use of rDNA techniques may reduce the widespread use of steroids, antibiotics, and related agents in animal feeds.

2. *Recombinant DNA techniques applied to control of environmental pollution*

Recombinant DNA micro-organisms have much to offer in efforts to minimise or overcome major problems of environmental pollution. Here organisms must be found or modified to meet demanding requirements of extreme environments, and they must survive and proliferate in competition with the existing flora, predators, and sometimes extreme or widely fluctuating chemical and physical conditions.

Waste streams from agriculture, forestry, industry, and municipal activities generate tremendous burdens on our environment. Micro-organisms, especially those indigenous to soils and landfills and agricultural and forestry waste sites, have remarkable degradative capabilities, though the level of these enzymatic activities is usually rather low. Although there is often extensive experience with their use, organisms which can degrade persistent and toxic chemicals or function in temperature or salinity extremes have been studied rather little, either physiologically or genetically. Furthermore, tools for genetic manipulation may be little developed in these species. On the other hand, *Pseudomonas* species, often found in natural sites, are under intensive study and have emerged as very promising agents for environmental applications.

Biological systems have been employed successfully for many years in both industrial and domestic pollution control. These are essentially chemostatic systems containing mixed populations capable of a wide range of biochemical functions. They must accept and efficiently mineralize relatively dilute and time-variant waste streams. The most promising prospects in rDNA organisms relate directly to manufacturing process modifications with simpler, consistent, and concentrated waste streams at the point of generation. Biochemical capacities most useful are dehalogenation, deamination, denitration, and ring-cleavage. Even partial detoxification of hazardous compounds would be a useful step in overall disposal strategies.

With regard to municipal wastes, microbiological means of reducing volume and bulk would be valuable.

One of the most challenging environmental applications of rDNA organisms is in degradation of highly toxic wastes. A considerable degree of research using rDNA organisms in this area is now under way. The literature contains numerous citations of uses of rDNA micro-organisms to degrade toxic compounds such as chlorinated phenols, cyanide, and dioxins.

However, in many instances the hazardous wastes can be toxic to the micro-organisms themselves, though proper selection techniques may lessen this problem. The earliest developments will probably involve enhanced degradation of toxic wastes in well-controlled environments (e.g., a well-characterised waste contained in a tank). Meanwhile, research on using rDNA micro-organisms to reduce concentrations of wastes in settling ponds or to clean up spills is also advancing. Decontamination of soils, ship hulls, and oil sumps are other good targets. Indigenous organisms could be endowed with much greater capacity for specific degradation of major toxic substrates by rDNA techniques.

3. *Microbial metal extraction and recovery*

Micro-organisms have been commercially used in the recovery of metals from ore for over a century. In the 1950s it was shown that *Thiobacillus ferroxidans* and *T. thiooxidans* oxidise some metals as well as several copper sulphide minerals. These micro-organisms are used today to leach copper and uranium from ores in significant commercial quantities. Similarly, work on using micro-organisms to concentrate metals for subsequent recovery has been developing since the 1960s. Organisms suggested for use include algae (e.g. *Chlorella vulgaris, Hormidium fluitans*), bacteria (e.g. *Pseudomonas aeruginosa, Bacillus subtilis, Escherichia coli*) and fungi (e.g. *Saccharomyces cerevisiae, Aspergillus niger*).

Many ore deposits have been mined to the point where the amount of metal which can be recovered using conventional techniques is no longer commercially feasible. Microbial leaching techniques can be used to obtain metal from these low-grade ores relatively cheaply. Also, microbial operations might be developed to be less energy intensive than other conventional recovery techniques. The benefits from use of rDNA techniques could be to develop organisms with greater tolerance to acidic and saline conditions, increased ability to survive high and low temperatures, and a capacity for leaching a greater number of metals.

4. *Enhanced oil recovery*

A considerable amount of the world's oil reserves remain in subterranean wells where the oil is either trapped in the rock formation or is too viscous to pump. Obstacles such as these have encouraged researchers to develop both speciality chemicals and microbial processes to recover more of the oil. A microbial biopolymer which is an effective hydrocarbon emulsifier, produced by *Acinetobacter calcoaceticus* modified by classical mutation and selection techniques, can be used to enhance removal of residual oil from tanker holds. This biopolymer is expected to undergo testing as an oil recovery enhancer which would help make the oil easier to pump from a well. It is possible that rDNA techniques may be useful in modifying other micro-organisms to improve or provide them with the ability to produce biopolymers specifically designed to maximise oil recovery rates.

Development of improved micro-organisms for direct injection into oil wells could also result from use of rDNA techniques. For a micro-organism to be commercially useful it must be able to survive severe heat, salinity and pressure conditions. Once in the well the micro-organism would either produce a gas to repressurise the well or produce surfactants or emulsifiers to lower the viscosity of the oil. Use of rDNA techniques to produce these micro-organisms could have a significant effect on increasing the amount of the world's recoverable oil reserves.

NOTES AND REFERENCES

1. *Commercial Biotechnology, An International Analysis*, Office of Technology Assessment, U.S. Congress, U.S. Government Printing Office, Washington, D.C., 1984.

2. *Health Impact of Biotechnology*, Report of a WHO (World Health Organisation) Working Group, Dublin, 9th-12th November, 1982, WHO, Copenhagen, 1984.

3. E. W. Houwink, *A Realistic View on Biotechnology*, Dechema, Frankfurt am Main, 1984 (published on behalf of the European Federation of Biotechnology).

4. Alan T. Bull, Geoffrey Holt, Malcolm D. Lilly, *Biotechnology – International Trends and Perspectives*, OECD, Paris, 1982.

5. I+ designates ice-nucleation positive or proficient
   I– designates ice-nucleation negative or deficient.

Chapter II

SAFETY CONSIDERATIONS

The purpose of this chapter is to propose a range of scientific considerations to be taken into account when assessing potential risks of industrial and environmental applications of micro-organisms, plants and animals and selecting appropriate safety measures.

Micro-organisms are associated with man, animals, plants and the environment, and their influence is often beneficial (see Chapter I). However, scientific considerations for safe use of these micro-organisms are needed as some of them are pathogenic and others may have a negative impact on the environment.

The potential applications of both conventional and modern biotechnology encompass a wide spectrum. They range from well-contained industrial uses of genetically-altered micro-organisms with specific traits or which synthesize desired products, to organisms intended for uncontained applications in agriculture or the environment. A number of existing scientific methods can be used to assess the potential risks associated with such applications.

Risk Assessment Methods

A detailed discussion of risk assessment techniques was considered to be beyond the scope of this report. This section contains a brief review of methods which are particularly relevant to the applications of rDNA micro-organisms. However they also apply in principle to plants and animals. The six following paragraphs have been adopted with minor modifications from a recent US report[1].

A useful descriptive model for biotechnology applications of micro-organisms was presented in a report by the US Office of Technology Assessment (OTA)[2]. The stages in OTA's model are the following:

1. *Formation* – the creation of a genetically-altered micro-organism through deliberate or accidental means.

2. *Release* – the deliberate release or accidental escape of some of these micro-organisms in the workplace and/or into the environment.

3. *Proliferation* – the subsequent multiplication, genetic reconstruction, growth, transport, modification and die-off of these micro-organisms in the environment, including possible transfer of genetic material to other micro-organisms.

4. *Establishment* – the establishment of these micro-organisms within an ecosystem niche, including possible colonisation of humans or other biota.

5. *Effect* – the subsequent occurrence of human or ecological effects due to interaction of the organism with some host or environmental factor.

The first two stages, formation and release, correspond to risk-source characterisation. These stages can, in principle, be analysed by quantifying the probabilities associated with the magnitudes of consequences. Estimation of probabilities and magnitudes is typically accomplished through fault-tree or event-tree analysis, or through simulations, as in a recent study by the Environmental Protection Agency (EPA)[3].

The last stage, human and ecological effects can, in principle, be analysed by adapting conventional epidemiological or toxicological methods. Using such methods, a number of studies have already been conducted to assess the potential for infection and disease activation in host organisms[4]. Another extensively-studied area has been the spread of antibiotic-resistant strains of bacteria in human populations. Thus, existing risk assessment methods are currently being used to characterise the consequences of human exposure to biotechnology products. Ecological consequence assessment, however, is a less well-developed field, and requires further research, although predictive modeling has been approached in epidemiological studies of agroecosystems.

The intermediate stages 3 and 4 are difficult to analyse using existing risk assessment methods. A key difficulty is the assessment of interactions of the micro-organism with the existing ecosystem. For example, an introduced micro-organism could transfer genetic material to other micro-organisms. Once established, micro-organisms can potentially alter the environment in ways that promote further proliferation or genetic transfer, giving rise to secondary effects.

Two important areas of investigation related to proliferation and establishment are: (i) environmental transport and fate of micro-organisms, and (ii) ecosystem interactions. Knowledge of micro-organism transport and fate (or survival) is useful for assessing potential exposures of non-target areas or non-target organisms. In recent years, scientific studies of transport and fate pathways have been conducted for a variety of bacteria, algae, viruses and other micro-organisms, in conjunction with studies of wastewater treatment processes, the spread of infectious agents, pesticidal applications, and water supply protection. Much of what has been learned in these studies is relevant in evaluating the transport and fate of genetically-altered micro-organisms.

Potential ecosystem interactions between genetically-modified micro-organisms and other existing organisms are extremely difficult to describe or predict accurately. One approach, qualitative risk assessment, can be used to compare the propensities for survival, establishment, and genetic stability under different environmental conditions. Proliferation of the introduced or subsequently altered micro-organisms might conceivably follow one of three patterns: temporary survival and eventual disappearance, establishment of one or several stable populations, or growth until limiting boundaries are reached. Although these principles are discussed in the context of micro-organisms, they also apply to plants and animals.

Considerations on Risk Assessment for rDNA Organisms

If one were to enhance or elicit a characteristic of an organism by conventional mutation or selection techniques, the safety assessment of the new organism would rely heavily on the knowledge of its parental organism, as well as on an analysis of how the new organism appears to differ from the parent. In the same way, any safety assessment of organisms altered by rDNA techniques designed for use in either an industrial, an agricultural or environmental application describes initially the relevant properties of the organism.

Recombinant DNA organisms are typically constructed by introducing a small segment of DNA from a "donor" organism into a "recipient" organism. The genome of the resulting organism derived by a rDNA technique from these two "parents" is therefore most like that of the recipient organism. Since all but a small fraction of the genetic information in the modified organism is that of the recipient organism, a description of the recipient's properties provides initial information useful in assessing the properties of the organism derived by rDNA techniques. Information describing the difference between the properties of the modified organism and of the recipient organism defines the framework for safety assessment.

If as a result of risk assessment it is concluded that the modified organism must be physically contained, this is feasible in the case of industrial use but not for agricultural or environmental applications. This distinction is even more fundamental than whether the modified organism has been derived by genetic manipulation or not.

1. Properties of donor and recipient organisms

Properties of the "recipient" organism that should be taken into account include: origin and classification, as well as genetic, pathogenic, physiological and ecological characteristics. Properties of the "donor" organism relate to the structure and function of the DNA sequences to be added. More information concerning the properties of the donor organism will be needed, when these DNA sequences are not fully characterised. While these kinds of information are generally developed during laboratory, field testing, and pilot plant stages, there may be cases where additional testing is necessary to generate additional data. (See Appendix B).

When chemically synthesized nucleic acids are used for construction of organisms, consideration may need to be given to the structure and function of those sequences.

2. Recombinant DNA technique for deriving the organism

The relevant properties of the recipient organism and the donor DNA provide information on the properties peculiar to the modified organism.

Description of the rDNA technique for deriving the organism provides important information on its anticipated properties. Component parts, for example, would include the donor nucleic acids, control elements, linking sequences, antibiotic-resistance genes, flanking regions and the like.

3. Properties of the organism derived by rDNA techniques

The essence of the assessment is the extent to which the recipient's properties are altered by the introduced DNA. A first consideration should be the degree of expression of the introduced genetic material. A second would be the extent to which relevant properties of the recipient have been modified as a result of the genetic manipulation, including significant new or unexpected effects. The close similarity between the recipient and modified organism allows tests useful in describing the properties of the modified organism.

Recombinant DNA techniques can be used to modify the genome of an organism, e.g. to delete a portion of a recipient genome. Compared to other kinds of manipulations, the use of a deletion technique would ordinarily suggest lesser concern about safety since a deletion typically makes smaller and more precisely defined changes, while also typically enfeebling the organism, and no new genetic information has been added to the parental organism. Deletions are also likely to mimic mutations that occur in organisms naturally. However, appropriate consideration should be given to the possibility of the expression of unanticipated functions particularly in the case of other types of modifications.

Safety Considerations Associated with Large-scale Industrial Applications

Traditional biological processes used in industry employ micro-organisms which are well characterised. For organisms considered to be of low risk, only minimal controls and containment procedures need to be used. With few exceptions (e.g. vaccine production) only organisms non-pathogenic to humans and animals are employed.

Conjectural hazards and evidence for safety[5]

When rDNA techniques were first introduced there was a natural concern as to their potential hazards, but after more than a decade of experimentation under controlled conditions, these hazards have remained conjectural and not based on incident. Initial uncertainty about safety of these techniques has been ameliorated by three compelling lines of evidence. First, experimental risk assessment studies specifically designed to test the hypothesis that host organisms can acquire unexpected hazardous properties from DNA donor cells have failed to demonstrate the existence of such conjectured hazards. Second, more rigorous evaluation of existing information regarding basic immunology, pathogenicity and infectious disease processes has resulted in the relaxation of containment specifications recommended by national authorities. Third, the experimentation conducted in recent years has elicited no observable novel hazard.

The above evidence suggests that the level of safety of micro-organisms derived by rDNA techniques may be evaluated reliably by examining the known properties of the components used in the recombinant DNA process. For example, when DNA coding for highly potential toxins is to be cloned, special attention is warranted.

Any potential hazards of industrial use of rDNA organisms are expected to be of the same nature as for other biological agents, namely:

i) Infection hazard – the potential for disease in man, animals and plants following exposure to the living organism or virus;

ii) The toxic, allergenic or other biological effect of the non-viable organism or cell, its components or its naturally occurring metabolic products;

iii) The toxic, allergenic or other biological effect of the product expressed by the organism;

iv) Environmental effects (see following section).

The development of appropriate practices and control measures has enabled biotechnology to be generally regarded as a safe industry. Where the conjectured hazards above *(i-iv)* exist, the risk of exposure needs to be addressed and adequate and appropriate measures taken to prevent or minimise such exposure. It should be said that relative to laboratory-scale work there is nothing intrinsically more hazardous about rDNA organisms or their products when industrial scale work is contemplated. It is mainly the scale of operation and hence the possible escape volume, concentration, and the duration of exposure that have increased. In addition, only under controlled fermentation conditions in a well-defined process will the biomass and the level of product per unit volume be maximised. This has to be balanced against the fact that most of the uncertainties related to the organism at the laboratory research stage have been eliminated prior to scale-up. Furthermore, disabled laboratory host strains can be used for fermentation production, provided that the organism is efficient under the process conditions determined within the fermenter. It should also be emphasized that there is an inherent incentive for industry to use organisms that pose a low risk. Not only does this minimise any national regulatory constraints, but it also involves lower costs by, for example, minimising the need for expensive plant containment and associated high containment safety procedures.

Further analysis of potential risks to humans, animals and plants of traditional and of recombinant organisms is given in Appendix E. These include pathogenicity, handling of bulk quantities of micro-organisms and consideration of biologically active products.

Safety Considerations Associated with Agricultural and Environmental Applications

1. General considerations

This section discusses the safety and hypothetical risks associated with the use of rDNA modified micro-organisms, plants, and animals in agricultural and environmental applications. The intentional genetic manipulation of living organisms dates from the recognition that organisms could be selected and crossbred to produce varieties of enhanced utility for such applications. Using rDNA techniques, very specific modifications can now be introduced into organisms, and barriers that have previously restricted the transfer of genetic material between species can be overcome or circumvented. Some of the micro-organisms, plants, and animals produced by rDNA techniques may, thus, differ qualitatively or quantitatively from the variants found in nature or developed through conventional breeding activities.

Concern has been expressed that application of these rDNA organisms in the environment may present ecological risks, and attempts have been made to evaluate this potential for harm. At this time, these attempts rest primarily on extrapolations from experiences with: (i) the introduction of naturally occurring organisms to ecosystems to which they are not native; (ii) evolution of novel traits in existing populations; and (iii) manipulation of agricultural crops, plant-associated microbes and animals. However our predictive knowledge in some of these areas is not comparable to that developed for industrial applications and consequently these areas should be kept under review as the field develops.

Past experience with species introductions have been studied in attempts to establish possible risk. In the great majority of instances no adverse consequences were noted. In some cases, however, introductions have produced biological changes in the receiving environments, a few of which were significant. It is impossible to know, moreover, the number of introductions which have failed to become established and are thus not noted in the literature.

Experience with evolution of "novel" traits in existing populations forms a basis for establishing parameters of potential risk. Genetic changes in species occur continually in nature. Occasionally a new trait may appear that confers a selective advantage resulting in an organism that could become more numerous, have an expanded host and geographical range, and/or utilise new resources and habitats. Although these events have been observed rarely, this experience indicates that a small genetic change may have a significant phenotypic effect. Depending on the nature of the change, the effects may be amplified in particular ecological settings such that the environmental impact may be significant.

Speculation that a small fraction of rDNA organisms could have deleterious effects on the environment is based on the following additional arguments: (i) the number of organisms to be used in some applications may be large; (ii) living organisms have the capacity to reproduce and spread in the environment; (iii) nucleic acid added to modify organisms to impart desired characteristics, might be transferred via plasmids, viruses, or other means to other organisms, which may then possess undesirable characteristics; (iv) rDNA techniques could produce organisms with different combinations from those that commonly occur in nature.

Any potential environmental impacts of agricultural and environmental applications of rDNA organisms are expected to be similar to effects that have been observed with introductions of naturally occurring species or selected species used for agricultural applications. They include: *(i)* direct but unanticipated effects of modified organisms on non-target species; *(ii)* effects on the outcome of direct interactions among species; *(iii)* alteration of indirect relationships between species; *(iv)* influences on the biochemical processes that support all ecosystems; and *(v)* changes in the rate and direction of the evolutionary responses of species to each other and to their physical and chemical environments.

As research and development on environmental and agricultural products progress, it is becoming apparent that rDNA techniques are being used to accomplish a number of objectives. These range from deletion of a small portion of an organism's genome, to greater control of a pathway already existing in an organism, to combinations of genetic material from closely related organisms, and to transfers of genes among very dissimilar organisms, producing combinations of traits not likely to occur in nature. Given this wide range of manipulations, it is likely that rDNA organisms and applications will fall into different risk categories, some of which will entail less risk than others. Some categorization of low-risk applications has been accomplished already[6].

The probability of unexpected adverse impacts from applications of rDNA modified organisms is often defined as low for several reasons. Recombinant DNA techniques are expected in most cases to produce better characterised organisms than those developed through traditional techniques. Success in developing useful engineered organisms depends on: selecting the appropriate recipient; identifying the gene(s) responsible for the desired function; isolating, cloning and transferring the gene(s) into the desired recipient; and regulating the introduced gene(s) to function as desired. Thus the construction of a useful organism requires a great deal of knowledge about the organism and the process of construction yields additional information.

The process of progressively decreasing physical containment, by which micro-organisms and higher organisms are developed routinely for agricultural and environmental applications – i.e., research in the laboratory, research in microcosms and other contained environments, small field testing, and large field testing, allows a logical, incremental step-wise process whereby safety and performance data are collected. In this development process, the organism is characterised and carefully observed at each stage and a prediction can be made of its behaviour in subsequent less confined stages of development.

2. *Considerations specific to micro-organisms*

Concerns about micro-organisms derived by rDNA techniques released into the environment include the possibility that the genetic modification might affect their host range, affect their capacity to utilise substrates such as nitrogen or lignin, convert them into pathogens, and/or alter the balance between them and ecologically interrelated populations in the ecosystem.

Since pathogenic organisms have the most obvious impact on human and agricultural systems, a major concern is the possibility that genetic modifications might convert non-pathogenic organisms to pathogens or alter the host range or virulence of pathogens used to control plant, insect, or other pests. Studies with pathogens have demonstrated that many genes must interact appropriately for a microbe to cause disease. The pathogen must possess and appropriately express characteristics such as recognition factors, adhesion ability, toxigenicity, and resistance to the host defence system. Single gene modifications of

29

organisms with no pathogenic potential or history, or introduction of several genes not contributing to pathogenicity, do not appear likely to result in unanticipated pathogenicity. Moreover, wide experience and extensive data on pathogenicity exist which can be used to define the parameters of concern when considering the effects of rDNA modifications.

3. *Considerations specific to plants*

Fewer concerns about conjectural risks have been advanced for agricultural applications of plants modified by rDNA techniques than have been advanced for microbes, because plants have generally been easier to monitor and control. Nonetheless, a specific concern is that rDNA of plants may produce a weed which will be difficult to control. The magnitude of concern would depend on the nature of the plant, the modified gene, and the environment into which the modified plant was to be introduced. Hypothetically, a weed might be produced: *(i)* inadvertently; *(ii)* through deliberate attempts to introduce hardier traits into crop plants; or *(iii)* through naturally occurring hybridization between wild plants and rDNA varieties. Hypothetically hybridization might transfer new genes to wild plants and introduce traits such as herbicide resistance, stress tolerance and insect resistance. Traits coded by a single gene whose allele is already present in many plants might have the highest probability of being transferred (e.g., herbicide resistance).

Taxonomic and genetic knowledge of weed genera and species suggest that the potential for producing weedy plants is greatest when weedy plants or plants belonging to a genus known to contain weeds, are manipulated for the development of new plant varieties. Yet, we have experience and success in this area through conventional plant breeding, e.g. improved tomato varieties. It is likely that a large number of genes must interact appropriately for a plant to display the properties of a weed, and this makes it unlikely that rDNA of plants will inadvertently develop a weedy plant. In any event, the chances of introducing "weediness" into a crop by rDNA techniques is far less likely than the introduction of such a characteristic by conventional plant breeding methods, in which weeds are often used as a source of genetic material for desirable traits such as disease and insect resistance.

A second potential concern in developing rDNA plants for human or animal consumption is the possibility that the modified plant may produce a toxic secondary metabolite or protein toxin, particularly if the plant is engineered for resistance to an insect pest. The same problem also arises in traditional plant breeding.

Some additional concerns in traditional plant variety development apply equally to any plant variety developed using rDNA techniques. These concerns include changes in plant-associated microflora in response to genetic changes in plants, or from growing a single plant variety intensively over a wide area. One example of the potentially adverse effects of such practices is the genetic response of pathogens and insects to changes in plant resistance. These responses may allow the pests to overcome plant resistance. This evolutionary response has been observed over time with a number of pests, and should not be considered a problem unique to rDNA applications.

4. *Considerations specific to animals*

Current experience with specific genetic manipulations of animals is limited, but the risks are expected to be low. The primary concerns associated with genetic modification of domesticated animals are: alterations in the regulation of the animal's genes, expression of endogenous latent viruses or exogenous genes, and the presence of unacceptable levels of bioactive materials in the food products. For aquatic or other animals which have or may have unconfined access to the environment, safety principles analogous to those for plants and microbes also apply. These possible problems have not been considered in detail in this report.

5. *Concluding remarks*

Safety concerns focus on whether environmental and agricultural applications of organisms modified by rDNA techniques pose an "incremental" risk. While rDNA techniques may result in the production of organisms expressing a combination of traits that are not observed in nature, genetic changes from rDNA techniques will often have inherently greater predictability compared to traditional techniques, because of the greater precision that the rDNA technique affords to particular modifications. It is expected that any risks associated with applications of rDNA organisms may be assessed in generally the same way as those associated with non-rDNA organisms. It is acknowledged that additional research and experience with rDNA micro-organisms, plants, and animals, should certainly increase our ability and precision to predict the outcome of introductions of rDNA organisms into the many varied ecosystems.

NOTES AND REFERENCES

1. *The suitability and Applicability of Risk Assessment Methods for Environmental Applications of Biotechnology*, eds: V.T. Covello and J.R. Fiksel, National Science Foundation, 1985, no. NSF/PRA 8502286.

2. *Impacts of Applied Genetics: Micro-organisms, Plants and Animals*, US Congress, Office of Technology Assessment, Washington, D.C., 1981.

3. Lincoln *et al., Release and Containment of Organisms from Applied Genetics Activities*, EPA Report, Carnegie-Mellon Univ., December 1983.

4. Summers and Kawanishi, *EPA Symposium, EPA Report no. 600/9-78 026*, Washington, D.C., 1978.

5. *Health Impact of Biotechnology*, Report of a WHO (World Health Organisation) Working Group, Dublin, 9th-12th November, 1982, WHO, Copenhagen, 1984; *Swiss Biotech*, no. 5, pp. 25-26 (1984).

6. See for example Appendix L of the US NIH Guidelines and the exemptions in Section 2.6 of the Australian Recombinant DNA Advisory Committee document: *The Planned Release of Live Organisms Modified by Recombinant DNA Techniques*, Interim and Consultation Edition, Department of Industry, Technology and Commerce, Canberra Act 2600, May 1985.

Chapter III

INDUSTRIAL LARGE-SCALE APPLICATIONS

This chapter addresses general principles for achieving safety in the large-scale industrial use of organisms manipulated with recombinant DNA techniques. In most OECD Member countries, large-scale industrial processes are considered to be those involving volumes greater than 10 litres of cultured organisms or metazoan cells. This quantitative distinction between laboratory scale and large scale is an arbitrary one and other volumes could be considered equally appropriate.

The vast majority of organisms now used in traditional manufacturing industry can be regarded as safe because in the course of long periods of industrial use, sometimes extending to centuries, they have rarely given rise to safety problems.

In the same way, modified organisms prepared by inserting segments of DNA that are well characterised and free from known harmful sequences into such organisms to improve their performance, are also unlikely to pose any risk.

Cases in which traditionally safe micro-organisms are modified by inserting segments of DNA to facilitate the manufacture of new products do not raise any safety considerations beyond those that might be posed by the products themselves. Each case should be assessed separately and in those few that are found to pose safety problems, production should be carried out under appropriate conditions of containment.

The purpose of such containment is to reduce exposure of workers and other persons, to prevent release of potentially hazardous agents into the outside environment, and to protect the product. It can be achieved either by "biological containment" (through exploiting natural barriers which limit an organism's ability to survive and/or transfer genetic information into specific environments) or by "physical containment".

Methods of containment are defined in more detail in Appendix G. Their use would be appropriate in the event that pathogenic organisms were used or that genes coding for harmful products were inserted.

Principles of Containment

Industrial safety programmes rely on two approaches: *(a)* biological containment, and *(b)* physical containment.

a) *Biological containment*

Natural barriers exist which limit an organism's ability to survive and/or transfer genetic information in specific environments. These highly specific barriers can be employed to aid in containing the organism. They include characteristics such as auxotrophy, UV sensitivity, etc.

When such barriers exist naturally or are introduced specifically into the organism, the organism is said to possess some degree of biological containment.

The degree of biological containment possessed by an organism can be influenced through manipulation of the organism or the vector. The most commonly employed biological containment modifications limit either:

   i)   The survival and multiplication of the organism in the environment; and/or

   ii)   The transmission of the genetic information to other organisms.

b)   *Physical containment*

The term "physical containment" includes three elements of containment: *(i)* equipment, *(ii)* operating practices and techniques, and *(iii)* facility design. Primary containment, i.e. the protection of personnel and the immediate vicinity of the process from exposure to these agents, is provided by appropriate equipment and the use of safe operating procedures. Secondary containment, i.e. the protection of the environment external to the facility from exposure to these materials, is provided by a combination of facility (building) design and operating practices:

i)   *Equipment*

–   The fermentation equipment used for industrial production utilising rDNA organisms serves as the principal means for achieving physical containment. The design of this equipment will vary according to the processes involved, size of vessels, and other such factors. The effectiveness of the primary containment provided by equipment is maintained by the complementary use of practices and procedures;

–   Primary containment will to a varying degree be provided by the plant itself. Additional primary containment may be achieved by, for example, exhaust-ventilated enclosures around potential leakage points;

ii)   *Operating practices and techniques*

–   An important element of containment is strict adherence to standard operating practices and techniques. Persons working with potentially infectious agents, allergenic or toxic materials, should be aware of potential hazards and should be trained and proficient in the practices and techniques for safely handling such material. The director or person in charge of the industrial facility should be responsible for providing or arranging for appropriate training of personnel;

–   When standard operating procedures are not sufficient to control a hazard, additional safety practices would then be selected. These additional practices should be in keeping with the agent or the procedure;

–   Some elements of operating procedure include: *(i)* a biosafety or operations manual which specifies the practices and procedures designed to minimise or eliminate risks; *(ii)* mechanisms to advise personnel of special hazards and to require personnel to read and to follow the required practices and procedures; *(iii)* direction of activities by an individual trained and knowledgeable in appropriate operating procedures, safety procedures, and any potential hazards in the workplace;

–   Personnel safety practices and techniques should be supplemented by appropriate facility design and engineering features, safety equipment, and management practices;

iii) *Facility design*

  — The design of the facility contributes to providing protection to the environment and persons outside the immediate production area. The degree of sophistication of design should be compatible with the production activities. The principles employed in designing various types of laboratories can be utilized in designing industrial facilities to achieve functional secondary barrier systems. It should be recognized that facility design is not independent of practices and procedures and primary containment devices in achieving containment.

Implementation of Containment

The primary objective in selecting containment is to match an appropriate level of physical measures and associated safety procedures to the conclusions of the risk assessment. The principles of containment represent an attempt to describe the end to be achieved rather than an attempt to specify the technical means of implementation. It is suggested that, for Good Industrial Large-Scale Practice (see below) as well as all levels of containment, the following fundamental principles of good occupational safety and hygiene be applied:

  i)   To keep workplace and environmental exposure to any physical, chemical or biological agent to the lowest practicable level;
  ii) To exercise engineering control measures at source and to supplement these with appropriate personal protective clothing and equipment when necessary;
  iii) To test adequately and maintain control measures and equipment;
  iv) To test when necessary for the presence of viable process organisms outside the primary physical containment;
  v)   To provide training of personnel;
  vi) To establish biological safety committees or subcommittees as required;
  vii) To formulate and implement local codes of practice for the safety of personnel.

Recombinant DNA-containing as well as other organisms used in industry will generally have been developed in the laboratory under a level of containment specified by guidelines governing research. The level of containment in the laboratory is one factor to be taken into account when assigning the appropriate level of containment for large-scale production.

Good Industrial Large Scale Practice (GILSP)

As described in this report and elsewhere[1], the hazards associated with rDNA micro-organisms can be assessed and managed in a similar way to those associated with other organisms. It should be recognized that, for organisms considered to be of low risk, only minimal controls and containment procedures are necessary. This will be the case for the vast majority of rDNA organisms used in industrial large scale production. For these reasons, we endorse the concept of Good Industrial Large-Scale Practice (GILSP) for organisms which may be handled at a minimal level of control. This would coincide, for example, with the controls recommended by the European Federation of Biotechnology for Class 1 organisms.

For organisms manipulated by rDNA techniques, criteria for allowing use of GILSP (see Appendix F) can be identified for the parental (host) organism, for the rDNA-engineered organism, and for the vector/insert employed:

  — The host organism should be non-pathogenic; should not contain adventitious agents; and should have an extended history of safe industrial use, or have built-in

34

environmental limitations that permit optimum growth in the industrial setting but limited survival without adverse consequences in the environment.
- The rDNA-engineered organism should be non-pathogenic; should be as safe in the industrial setting as the host organism, and without adverse consequences in the environment.
- The vector/insert should be well-characterised and free from known harmful sequences; should be limited in size as much as possible to the DNA required to perform the intended function; should not increase the stability of the construct in the environment unless that is a requirement of the intended function; should be poorly mobilisable; and should not transfer any resistance markers to micro-organisms not known to acquire them naturally if such acquisition could compromise the use of a drug to control disease agents in human or veterinary medicine or agriculture.

There are two clear examples of other classes of organisms that warrant the GILSP designation *unless they are pathogenic*:

i) Those constructed entirely from a single prokaryotic host (including its indigenous plasmids and viruses) or from a single eukaryotic host (including its chloroplasts, mitochondria or plasmids – but excluding viruses –); and

ii) Those consisting entirely of DNA segments from different species that exchange DNA by known physiological processes.

Evaluation of Organisms Used in Industrial Processes Under Specified Levels of Physical Containment

It is recognized that there may be a need in some cases to use specified levels of physical containment. The properties of organisms which are of interest in the evaluation of potential risk include the organisms' biological containment and their potential adverse effects. Several categories of information may be considered in evaluating the potential for adverse effects of a rDNA organism. Firstly there are the characteristics of the donor and recipient organisms and the introduced DNA. Points to consider in evaluating such organisms and the introduced DNA are outlined in Chapter II and Appendix B of this report. They serve as an aide-memoire but are not necessarily applicable in all cases. The second category involves evaluation of the characteristics of the rDNA organisms themselves. Appendix C lists those considerations and others that are of particular relevance in evaluating the potential for human health effects in the industrial setting. Appendix D lists those characteristics of engineered organisms which might be considered as part of a contingency plan should a release from the facility occur. It should be remembered that Appendix D was developed primarily with environmental applications in mind, while most large-scale applications typically utilise organisms with limited ability to survive in the environment. A more limited application of the points to consider is appropriate when evaluating environmental impacts of rDNA organisms used in large-scale industrial applications (see Chapter IV). It is anticipated that much of the necessary information will have been developed during the laboratory and pilot plant stage of an industrial process.

Matching Physical Containment with the Assessment of Potential Risk

Long before rDNA techniques were developed, procedures were employed for physical containment for large-scale industrial processes. Recombinant DNA organisms can also be contained using these standard physical containment principles. Obviously the level of physical containment must match the assessment of the risk.

Other considerations will influence selection of industrial containment, however. These considerations are: *(i)* the nature of the modified organism; and *(ii)* the nature of the product and the industrial process. In some cases, an evaluation of the modified organism may indicate that the containment level appropriate for laboratory construction of the organism is not appropriate to the large-scale process. For example, the laboratory level of physical containment may be high if the donor organism is a pathogen; however, the resulting modified rDNA-containing organism may be a non-pathogen which contains donor DNA sequences not associated with the pathogenic phenotype (e.g., *E. coli* host-vector systems expressing hepatitis B surface antigen). Lower containment levels might then be appropriate for any laboratory studies subsequent to construction of this type of rDNA organism. Application of a lower physical containment level to the industrial process utilising this modified organism might also be appropriate. However, because some considerations associated with the industrial process may differ from those associated with laboratory research, the modified organism should be re-evaluated and appropriate containment selected at the time of transfer to large-scale processes.

It should also be recognized that in some cases, the risks presented by other aspects of the process and by the product may dictate the level of physical containment. Industrial process plants and equipment are more diverse in application and scale than the typical research laboratory; thus, the methods selected for the physical control of risks will be more diverse. In addition, the industrial processes will probably have to be considered in unit process steps. The requirements of the particular portion of the process will dictate the physical containment to be used in that portion. This procedure will permit the degree of latitude required by the great diversity of industrial settings, and allow selection of procedures and design best fitted to assure adequate and safe containment.

As long as these methods provide the required containment, flexibility in selection is desirable. It may, thus, be appropriate to select and combine containment requirements from different categories on the basis of a unit process assessment. Therefore, it is not useful to describe fixed categories of containment. Examples of possible containment categories are given in Appendix G. Because of the rapid advance of knowledge, precise risk assessment as it relates to physical containment may be revised as experience accumulates.

NOTE AND REFERENCE

1. Küenzi *et al.,* "Safe Biotechnology – General Considerations", in: *Applied Microbiology and Biotechnology*, Springer-Verlag, 1985, Vol. 21, pp. 1-6. A report prepared by the Safety in Biotechnology Working Party of the European Federation of Biotechnology.

Chapter IV

ENVIRONMENTAL AND AGRICULTURAL APPLICATIONS

Recombinant DNA containing organisms will be developed for purposes involving uncontained applications in the environment, e.g. as pesticides, to improve plant growth, leach ore, enhance oil recovery and to degrade pollutants. It is likely too that rDNA plants and animals will be used to enhance production of fibre, food and fodder. Other applications are likely to include production and use of certain animal and human drugs. Release of rDNA organisms from industrial sites may also occur and should be considered as well.

In current agricultural research and development practices conventionally modified organisms are generally extensively tested prior to commercialisation. This testing typically involves a stepwise process that may range from greenhouse or other type of specialised containment to limited scale controlled field plots, and finally to large-scale multiple field plots in various geographic sites. Similar procedures are appropriate in the development of rDNA organisms for agricultural applications. Clinical applications of products intended for use in animals and humans also go through investigational trial stages of development prior to commercialisation. Environmental applications of micro-organisms are usually pre-tested in controlled systems prior to actual release. Evaluations and data collection at each incremental stage are conducted, both to ascertain efficacy and to eliminate any organism or application resulting in untoward environmental effects (or which is ineffective for the intended purpose). These data are often relevant to the evaluation of environmental effects.

It is not likely that a single set of scientific considerations will apply for all of the various organisms, environments, and release patterns envisioned for the agricultural and environmental applications of rDNA modified organisms. Rather, the particular application and the nature of the modified organism will indicate whether a safety assessment is warranted and which specific considerations are relevant. The discussion in this chapter is intended to describe the various scientific considerations that may be relevant in such an assessment, and is *not* intended to propose universally applicable regimes or standards for review of proposals. It is important that as environmental and agricultural applications of rDNA organisms proliferate, their safety record be compiled and compared to that of conventional organisms.

Considerations in Evaluating Safety of Environmental and Agricultural Applications of rDNA Organisms

The scientific considerations in this chapter are organised around a series of events, all of which must occur before an adverse effect can result. An evaluation of the safety of an organism is therefore structured around a consideration of these events in a step-wise

37

assessment. If the likelihood of any one of these events occurring is low, then the overall likelihood that any adverse environmental effect will occur will be low.

1. *Application in the environment*

The location and nature of the site of application, and the method and magnitude of the application are important for assessing safety. Agricultural applications for food, feed and fibre production may result in release of large quantities of modified organisms into terrestrial or aquatic ecosystems (e.g., herbicide-resistant plants, non-ice-nucleating bacteria, viral or bacterial pesticides, transgenic animals). Recombinant DNA-derived vaccines for animals and humans, as well as certain plant-associated micro-organisms, will have a much more limited pattern of environmental exposure because of biological specificity to the host, but incidental release to the environment certainly occurs in sewage and feed-lot or irrigation runoff waters, and may be significant. Environmental applications (e.g., metal extraction, pollutant and toxic waste degradation) may be confined initially to a specific location or may result in broad ecosystem exposure. The scientific considerations for assessing safety will vary with each particular application, depending on the organism, the physical and biological proximity to man and/or other significant biota. Local quarantine regulations and monitoring methodologies utilised during research and development will also be relevant.

2. *Survival, multiplication and/or dissemination in the environment*

The relative ability of the organism to survive and multiply in the environment in which it is applied and to be disseminated to new environments, is an important consideration for assessing the safety of the release. Undoubtedly, rDNA modified organisms used for agricultural and environmental applications will possess a certain capacity for survival and reproduction, and perhaps dissemination, in order to achieve the intended purpose of the application. This is in contrast to biologically debilitated (contained) organisms used for many large-scale industrial applications (see Chapter III). An assessment might attempt to determine whether the modified organism is likely to differ from the non-modified organism during exposure to relevant environmental conditions affecting survival and reproduction (e.g., climatic and soil factors); whether the modification affects the route and/or extent of dissemination of the organism; and whether an excessive increase in the number of organisms could occur and result in untoward environmental effects. In addition, preference should be given to the use of rDNA vectors which are limited in their ability to transfer into other organisms and thus spread in the environment.

3. *Interactions with species or biological systems*

Two general considerations are important in describing the interactions between the rDNA modified organism and the ecosystem to which it is applied: a description of the ecosystem (e.g., habitat, predominant species), and the potential interactions (e.g., pathogenicity, gene transfer, excessive growth in numbers) in that ecosystem. A complete description of the elements comprising an ecosystem is rarely possible; thus considerations should usually focus on the important features of the particular ecosystem.

The characteristics of the rDNA modified organism as compared to the original organism will serve as a guide in determining which interactions are most likely to be of significance. Points to consider in this evaluation are given earlier in Chapter II and in Appendices B and D.

4. *Effects on the environment*

The factors which may lead to significant environmental impacts are considered at this stage of assessment. These include: *(i)* effects on other organisms such as pathogenicity, infectivity and, effects on competitors, prey, hosts, symbionts, etc.; *(ii)* known or predicted involvement in biogeochemical processes such as mineral cycling, nitrogen fixation, etc.; *(iii)* genetic or phenotypic stability of the released organism; *(iv)* probability of transfer of genetic material to other organisms in the ecosystem; *(v)* the effect of excessive increase in numbers of organisms following the application.

Relevance to Plants and Animals

Many of the scientific considerations described in this chapter are relevant to plants and animals derived by rDNA techniques. Additionally, the general considerations in Chapter II describing the significance of the donor, recipient and modified organisms are also essential to the initial phase of the safety assessment.

There are major differences between the environmental safety assessment of plants and animals in comparison to micro-organisms. These arise primarily because of differences in complexity and size, as well as life span and degree of genetic isolation or biological containment. Concerns about detection, monitoring and containment relevant for applications of micro-organisms are not likely to arise to the same degree with plants and animals.

Availability of Information and Test Methodologies

The points in Appendix D illustrate the kinds of parameters that should be considered generally in evaluating the environmental impacts of releasing organisms. Though assessment of risks of introducing organisms into the environment is not a well-developed research field, there does exist a large body of information on ecology, pathology, taxonomy and physiology pertaining to microbes, plants, and animals which can be used as a source of data.

As the state-of-the-art involving applications of rDNA organisms progresses from research to field trials and commercial applications, it is expected that more scientific information will become available. Additional research to increase our ability to predict the outcome of introductions of rDNA organisms into ecosystems should be encouraged. The most complete data on survival and dissemination, as well as the best test methodologies, probably exist for those micro-organisms which are pathogenic to domestic plants and animals.

Existing data on micro-organisms have been collected, systematised and computerised to a limited extent. Strain-specific data systems are beginning to be co-ordinated internationally. Gene sequence data and functional maps of genetic elements (e.g. plasmids, phages) are being automated. Several viral genomes have been sequenced in their entirety.

Information derived from the long history of introducing, breeding and releasing domesticated plant and animal species and certain microbes can provide important baseline data for assessing the rDNA organisms. As noted earlier, information on the recipient organism may well be useful in predicting the fate of the modified organism.

Situations could arise where the concern about untoward environmental effects of an engineered organism might require additional data. While data from contained or simulated environments are often useful in assessment, the organism may not perform in the field as it performs in the simulated environments. Small-scale field tests may thus provide the only mechanism for obtaining valid data.

Evaluating Environmental Risks of rDNA Organisms Released from Industrial Applications

The risk assessment of large scale industrial applications should take into account the possible accidental or incidental release of organisms into the environment. The vast majority of large scale industrial applications utilise GILSP organisms which either have a long history of safe use and/or limited survival capacity outside the facility. The degree of physical containment, including emissions control, normally employed will be dictated by the biological properties of the modified organisms.

For large-scale industrial applications that result in significant release of viable organisms into the environment, an assessment of safety should use the scientific considerations described in this chapter.

Chapter V

SUMMARY AND RECOMMENDATIONS

SUMMARY OF MAJOR POINTS

Recombinant DNA techniques have opened up new and promising possibilities in a wide range of applications and can be expected to bring considerable benefits to mankind. They contribute in several ways to the improvement of human health and the extent of this contribution is expected to increase significantly in the near future.

The vast majority of industrial rDNA large-scale applications will use organisms of intrinsically low risk which warrant only minimal containment [GILSP (Good Industrial Large-Scale Practice)].

When it is necessary to use rDNA organisms of higher risk, additional criteria for risk assessment can be identified and furthermore, the technology of physical containment is well known to industry and has successfully been used to contain pathogenic organisms for many years. Therefore, rDNA micro-organisms of higher risk can also be handled safely under appropriate physical and/or biological containment.

Assessment of potential risks of organisms for environmental or agricultural applications is less developed than the assessment of potential risks for industrial applications. However, the means for assessing rDNA organisms can be approached by analogy with the existing data base gained from the extensive use of traditionally modified organisms in agriculture and the environment generally. With step-by-step assessment during the research and development process, the potential risk to the environment of the applications of rDNA organisms should be minimised.

RECOMMENDATIONS

I.   General

1.   Harmonization of approaches to rDNA techniques can be facilitated by exchanging: principles or guidelines for national regulations; developments in risk analysis; and practical experience in risk management. Therefore, information should be shared as freely as possible.

2.   There is no scientific basis for specific legislation for the implementation of rDNA techniques and applications. Member countries should examine their existing oversight and review mechanisms to ensure that adequate review and control may be applied while avoiding any undue burdens that may hamper technological developments in this field.

3.   Any approach to implementing guidelines should not impede future developments in rDNA techniques. International harmonization should recognise this need.

4.   To facilitate data exchange and minimise trade barriers between countries, further developments such as testing methods, equipment design, and knowledge of microbial taxonomy should be considered at both national and international levels. Due account should be taken of ongoing work on standards within international organisations, e.g. World Health Organisation (WHO); Commission of the European Communities (CEC); International Standards Organisation (ISO); Food and Agriculture Organisation (FAO); Microbial Strains Data Network (MSDN).

5.   Special efforts should be made to improve public understanding of the various aspects of rDNA techniques.

6.   For rDNA applications in industry, agriculture and the environment, it will be important for Member countries to watch the development of these techniques. For certain industrial applications and for environmental and agricultural applications of rDNA organisms, some countries may wish to have a notification scheme.

7.   Recognizing the need for innovation, it is important to consider appropriate means to protect intellectual property and confidentiality interests while assuring safety.

II.   Recommendations Specific for Industry

1.   The large-scale industrial application of rDNA techniques should wherever possible utilise micro-organisms that are intrinsically of low risk. Such micro-organisms can be handled under conditions of Good Industrial Large-Scale Practice (GILSP).

2.   If, following assessment using the criteria outlined in the report, a rDNA micro-organism cannot be handled merely by GILSP, measures of containment corresponding to the risk assessment should be used in addition to GILSP.

3.   Further research to improve techniques for monitoring and controlling non-intentional release of rDNA organisms should be encouraged in large-scale industrial applications requiring physical containment.

III.   Recommendations Specific for Environmental and Agricultural Applications

1.   Considerable data on the environmental and human health effects of living organisms exist and should be used to guide risk assessments.

2.   It is important to evaluate rDNA organisms for potential risk, prior to applications in agriculture and the environment. However, the development of general international guidelines governing such applications is premature at this time. An independent review of potential risks should be conducted on a case-by-case[1] basis, prior to the application.

3.   Development of organisms for agricultural or environmental applications should be conducted in a stepwise fashion, moving, where appropriate, from the laboratory to the growth chamber and greenhouse, to limited field testing and finally, to large-scale field testing.

4.   Further research to improve the prediction, evaluation, and monitoring of the outcome of applications of rDNA organisms should be encouraged.

NOTE AND REFERENCE

1.   Case-by-case means an individual review of a proposal against assessment criteria which are relevant to the particular proposal; this is not intended to imply that every case will require review by a national or other authority since various classes of proposals may be excluded.

APPENDICES

DEFINITIONS

Definition of genetic manipulation and Recombinant DNA varies in detail between countries. Examples include:

1. **United Kingdom** – Health and Safety (Genetic Manipulation) Regulations, 1978:

    " 'Genetic manipulation' means the formation of new combinations of heritable material by the insertion of nucleic acid molecules, produced by whatever means outside the cell, into any virus, bacterial plasmid, or other vector system so as to allow their incorporation into a host organism in which they do not naturally occur but in which they are capable of continued propagation".

2. **United States** – Guidelines for Research Involving Recombinant DNA Molecules, June 1983:

    "Definition of Recombinant DNA Molecules. In the context of these Guidelines, recombinant DNA molecules are defined as either *(i)* molecules which are constructed outside living cells by joining natural or synthetic DNA segments to DNA molecules that can replicate in a living cell, or *(ii)* DNA molecules that result from the replication of those described in *(i)* above. NOTE: Synthetic DNA segments likely to yield a potentially harmful polynucleotide or polypeptide (e.g. a toxin or a pharmacologically active agent) shall be considered as equivalent to their natural DNA counterpart. If the synthetic DNA segment is not expressed *in vivo* as a biologically active polynucleotide or polypeptide product, it is exempt from the Guidelines".

GENERAL SCIENTIFIC CONSIDERATIONS

This Appendix attempts to set out basic scientific considerations that may be relevant in considering the possible risks associated with the use of rDNA organisms. Although the list attempts to be comprehensive as far as present knowledge allows, not all the points included will apply to every case. It is to be expected therefore that individual proposals will address only the particular subset of issues that are appropriate to individual situations. The level of detail required in response to each subset of considerations is also likely to vary according to the nature of the proposed activity.

A. Characteristics of Donor and Recipient Organisms

1. *Taxonomy, identification, source, culture*

   a) Names and designations;
   b) The degree of relatedness between the donor and recipient organisms and evidence indicating exchange of genetic material by natural means;
   c) Characteristics of the organism which permit identification and the methods used to identify the organisms;
   d) Techniques employed in the laboratory and/or environment for detecting the presence of, and for monitoring, numbers of the organism;
   e) The sources of the organisms;
   f) Information on the recipient organism's reproductive cycle (sexual/asexual);
   g) Factors which might limit the reproduction, growth and survival of the recipient organism.

2. *Genetic characteristics of donor and recipient organisms*

   a) History of prior genetic manipulation;
   b) Characterisation of the recipient and donor genomes;
   c) Stability of recipient organism in terms of relevant genetic traits.

3. *Pathogenic and physiological traits of donor and recipient organisms*

   a) Nature of pathogenicity and virulence, infectivity, or toxigenicity;
   b) Host range;
   c) Other potentially significant physiological traits;
   d) Stability of these traits.

B. Character of the Engineered Organism

a) Description of the modification;

b) Description of the nature, function and source of the inserted donor nucleic acid, including regulatory or other elements affecting the function of the DNA and of the vector;

c) Description of the method(s) by which the vector with insert(s) has been constructed;

d) Description of methods for introducing the vector-insert into the recipient organism and the procedure for selection of the modified organism;

e) Description of the structure and amount of any vector and/or donor nucleic acid remaining in the final construction of the modified organism;

f) Characterisation of the site of modification of the recipient genome. Stability of the inserted DNA;

g) Frequency of mobilisation of inserted vector and/or genetic transfer capability;

h) Rate and level of expression of the introduced genetic material. Method and sensitivity of measurement;

i) Influence of the recipient organism on the activity of the foreign protein.

Appendix C

HUMAN HEALTH CONSIDERATIONS

This Appendix attempts to set out potential human health considerations associated with the use of rDNA organisms. Although the list provided attempts to be comprehensive as far as present knowledge allows, not all the points included will apply to every case. It is to be expected therefore that individual proposals will address only the particular subset of issues that are appropriate to individual situations. The level of detail required in response to each subset of considerations is also likely to vary according to the nature of the proposed activity.

A. Characteristics of the Engineered Organism

    1.   Comparison of the engineered organism to the recipient organism regarding pathogenicity.
    2.   Capacity for colonisation.
    3.   If the organism is pathogenic to humans (or to animals if appropriate):

        a)   Diseases caused and mechanism of pathogenicity including invasiveness and virulence;
        b)   Communicability;
        c)   Infective dose;
        d)   Host range, possibility of alteration;
        e)   Possibility of survival outside of human host;
        f)   Presence of vectors or means of dissemination;
        g)   Biological stability;
        h)   Antibiotic-resistance patterns;
        i)   Toxigenicity;
        j)   Allergenicity.

B. Health Considerations Generally Associated with the presence of Non-Viable Organisms or with the Products of rDNA processes

    1.   Toxic or allergenic effects of non-viable organisms and/or their metabolic products.
    2.   Product hazards.

C. Management of Personnel Exposure

    1.   Biological Measures:

        a)   Availability of appropriate prophylaxis and therapies;
        b)   Availability of medical surveillance.

    2.   Physical and organisational measures.

ENVIRONMENTAL AND AGRICULTURAL CONSIDERATIONS

This Appendix attempts to set out potential environmental and agricultural implications associated with the use of rDNA organisms. Although the list attempts to be comprehensive as far as present knowledge allows, not all the points included will apply to every case. It is to be expected therefore that individual proposals will address only the particular subset of issues that are appropriate to individual situations. The level of detail required in response to each subset of considerations is also likely to vary according to the nature of the proposed activity.

A. Ecological Traits relating to the Donor and Recipient

   a) Natural habitat and geographic distribution. Climatic characteristics of original habitats;
   b) Significant involvement in environmental processes;
   c) Pathogenicity – host range, infectivity, toxigenicity, virulence, vectors;
   d) Interactions with and effects on other organisms in the environment;
   e) Ability to form survival structure (e.g., seeds, spores, sclerotia);
   f) Frequency of genotypic and phenotypic change;
   g) The role of the genetic material to be donated on the ecology of the donor organism;
   h) The predicted effect of the donated genetic material on the recipient organism.

B. Application of the Engineered Organism in the Environment

   a) Geographical location of site, physical and biological proximity to man and/or any other significant biota;
   b) Description of site including size and preparation, climate, temperature, relative humidity, etc.;
   c) Containment and decontamination;
   d) Introduction protocols including quantity and frequency of application;
   e) Methods of site disturbance or cultivation;
   f) Methods for monitoring applications;
   g) Contingency plans;
   h) Treatment procedure of site at the completion of application.

C. Survival, Multiplication and Dissemination of the Engineered Organism in the Environment

   1. *Detection, identification and monitoring techniques*

   a) Description of detection, identification and monitoring techniques;
   b) Specificity, sensitivity and reliability of detection techniques;
   c) Techniques for detecting transfer of the donated DNA to other organisms.

2. *Characteristics affecting survival, multiplication and dissemination*

    a) Biological features which affect survival, multiplication or dissemination;

    b) Behaviour in simulated natural environments such as microcosms, growth rooms, greenhouses, insectaries, etc.;

    c) Known and predicted environmental conditions which may affect survival, multiplication, dissemination.

D. Interactions of Engineered Organism(s) with Biological Systems

1. *Target and non-target populations*

    a) Known and predicted habitats of the engineered organism;

    b) Description of the target ecosystems and of ecosystems to which the organism could be disseminated;

    c) Identification and description of target organisms;

    d) Anticipated mechanism and result of interaction between the engineered organism and the target organism(s);

    e) Identification and description of non-target organism(s) which might be exposed.

2. *Stability*

    a) Stability of the organism in terms of genetic traits;

    b) Genetic transfer capability;

    c) Likelihood of post-release selection leading to the expression of unexpected and undesirable traits by the engineered organism;

    d) Measures employed to ensure genetic stability, if any;

    e) Description of genetic traits which may prevent or minimise dispersal of genetic material.

3. *Routes of dissemination*

    a) Routes of dissemination, physical or biological;

    b) Known or potential modes of interaction, including inhalation, ingestion, surface contact, burrowing and injection.

E. Potential Environmental Impacts

1. *Potential effects on target and non-target organisms*

    a) Pathogenicity, infectivity, toxigenicity, virulence, vector of pathogen, allergenicity, colonisation;

    b) Known or predicted effects on other organisms in the environment;

    c) Likelihood of post-release shifts in biological interactions or in host range.

2. *Ecosystems effects*

    a) Known or predicted involvement in biogeochemical processes;

    b) Potential for excessive population increase.

Appendix E[1]

POTENTIAL RISKS TO HUMANS, PLANTS AND ANIMALS OF TRADITIONAL AND RECOMBINANT ORGANISMS

1. Pathogenicity

Pathogenicity is the potential ability of living organisms and viruses to cause disease in man, animals and plants. A small proportion of the very large number of micro-organisms known to science has this ability. Such diseases are the result of interactions between the parasite and the host, and it is not possible to conclude that a particular micro-organism or virus will cause disease because pathogenicity always depends on the genetic make-up and physiological state of both host and parasite as well as other factors, including the infecting dose and the portal of entry into the host.

Very few of the micro-organisms and viruses that are capable of causing disease are used in industry and these are usually necessarily employed in the manufacture of vaccines, toxoids or diagnostic reagents.

The safety problem facing a manufacturer who wishes to introduce a new technological process is to determine whether the organism on which it is based is capable of causing disease, and if it is, to decide an appropriate method of containment. In addressing this problem the manufacturer will first have the organism classified to genus and species. This will permit an initial assessment of its probable behaviour as a pathogen, based on existing knowledge of the organism itself and of related species. This assessment may then be supplemented by pathogenicity tests.

Clinical and research laboratories in which micro-organisms of all kinds, including pathogens, are handled routinely have adopted classifications which grade them into Risk Classes, usually 1-4 in increasing order of pathogenicity and on the basis of hazard to worker and general public.

It is useful to raise the possibility of a non-pathogenic cellular organism obtaining pathogenic characteristics after the introduction of a DNA insert from another organism. In the case of *E. coli* for example, the best characterised organism in molecular genetics, such an insert is likely to be approximately 10^{-3} times the size of the genome DNA. *E. coli* K12, an organism of choice for recombinant DNA work, was isolated some 50 years ago and during these years of laboratory maintenance has lost many characteristics associated with the wild-type *E. coli* including:
- The cell surface K antigen;
- Part of the lipopolysaccharide side chain;
- An adherence factor (*fimbriae*) which enabled the original strain to adhere to epithelial cells of the human gut;
- Resistance to lysis by complement in human serum; and
- Some resistance to phagocytosis by white blood cells.

Thus *E. coli* K12 does not colonise the human gut and is non-pathogenic. Four of the five genes specifying the above properties are widely separated on the *E. coli* chromosome. It is reasonable to conclude that the number of genes involved in pathogenicity is much too large for accidental transfer in a recombinant DNA experiment. Similar arguments are applicable to other micro-organisms (e.g. *B. subtilis* and *S. cerevisiae*) used in such experiments.

It could be argued that in a shotgun experiment even such an improbable event could occur. As indicated above, however, the results of experimentation conducted over a number of years do not support this conjecture. In addition, when a strain has been constructed for production, the selected DNA has been defined, sequenced and cut to the minimum required size. The properties of the product and the strain involved will have been studied in detail, and the production strain will differ from the parent only in its ability to produce the desired product. In the production unit itself, therefore, the risk to personnel of infection by a production strain that became pathogenic is negligible.

2. Safe Handling of Bulk Quantities of Micro-organisms

In some processes bulk quantities of micro-organisms remain after the desired product has been separated, and these require safe disposal. If they are pathogenic, they are killed by physical or chemical means or by a combination of such methods.

There then remains only a disposal problem, the nature of which is determined by the composition of the cell mass.

3. Safety of Biologically Active Products

It is with respect to the quality control of *products* that some of the conjectural hazards related to changes in the organism and quality of the product it produces should be considered on a case-by-case basis. It is already standard practice in the production of proteins for therapeutic use, employing rDNA techniques, to test the integrity of the vector and gene as well as the final product at regular intervals. This should be sufficient in most cases to detect biological changes in the process. For example, it is as part of the stringent quality control of products from animal cell culture that screening for the presence of virus (e.g. retrovirus) components and host DNA, which may contain 'hazardous' genetic material occurs. Therefore, the purification of the product from cellular protein and nucleic acid is required.

Potential product hazards from plant cells are those generally associated already with the pharmaceutical industry, and can be controlled by existing technology.

The potential hazards related to products are common whether the process involves microbial, animal or plant cells, and independent of rDNA techniques. These will be considered on a case-by-case basis. Product hazard will mainly arise by aerosol formation and escape into the environment. The following criteria should be considered:

  i) It is directly toxic to human, animal or plant life;
  ii) It can be absorbed by the nasal, eye tissue or lung alveoli or via the buccal cavity or by direct skin contact, particularly in the case of skin abrasions;
  iii) The product is converted by secondary metabolism into a toxic entity in the tissue which it invades;
  iv) The product causes an immune or allergenic response;
  v) The exposure is sufficient to elicit a toxic or an immune response.

In dealing with these biologically active microbial products, the biotechnology industries have an advantage in that their processes are carried out under controlled conditions which allow the detection of such products and permit their containment or removal to whatever standards are found to be necessary. Thus biotechnology using rDNA techniques is built upon standard industrial fermentation practices that have been used successfully for many years.

NOTES AND REFERENCES

1. Points 1 and 2 of this Appendix are partly derived from: *Health Impact of Biotechnology, op. cit.*; and "Safe Biotechnology – General Considerations", in: *Applied Microbiology and Biotechnology, op. cit.*

Appendix F

SUGGESTED CRITERIA FOR rDNA GILSP
(GOOD INDUSTRIAL LARGE SCALE PRACTICE) MICRO-ORGANISMS

Host Organism	rDNA Engineered Organism	Vector/Insert
– Non-pathogenic;	– Non-Pathogenic;	– Well characterised and free from known harmful sequences;
– No adventitious agents;	– As safe in industrial setting as host organism, but with limited survival without adverse consequences in environment.	– Limited in size as much as possible to the DNA required to perform the intended function; should not increase the stability of the construct in the environment (unless that is a requirement of the intended function);
– Extended history of safe industrial use; *OR*		– Should be poorly mobilis - able;
– Built-in environmental limitations permitting optimal growth in industrial settting but limited survival without adverse consequences in environment.		– Should not transfer any resistance markers to micro-organisms not known to acquire them naturally (if such acquisition could compromise use of drug to control disease agents).

EXAMPLES OF CONTAINMENT APPROACHES FOR LARGE SCALE INDUSTRIAL APPLICATIONS OTHER THAN GILSP (GOOD INDUSTRIAL LARGE SCALE PRACTICE)

A. Category 1

At this level of physical containment the following objectives should be achieved:

a) Viable organisms should be handled in a production system which physically separates the process from the environment;

b) Exhaust gases should be treated to minimise (i.e., to reduce to the lowest practicable level consistent with safety) the release of viable organisms;

c) Sample collection, addition of materials to the system and the transfer of viable organisms to another system should be done in a manner which minimises release;

d) Bulk quantities of culture fluids should not be removed from the system unless the viable organisms have been inactivated by validated means;

e) Closed systems should be located in an area controlled according to the requirements 6 *(c)* and *(d)* specified in the Table hereafter;

f) Effluent from the production facility should be inactivated by validated means prior to discharge.

B. Category 2

At this level of physical containment the following objectives should be achieved:

a) Viable organisms should be handled in a production system which physically separates the process from the environment;

b) Exhaust gases should be treated to prevent the release of viable organisms;

c) Sample collection, addition of materials to a closed system and the transfer of viable organisms to another closed system should be done in a manner which prevents release;

d) Culture fluids should not be removed from the closed system unless the viable organisms have been inactivated by validated chemical or physical means;

e) Seals should be designed to prevent leakage or should be fully enclosed in ventilated housings;

f) Closed systems should be located in an area controlled according to the requirements 6 *(a)*, *(b)*, *(c)*, and *(d)* specified in the Table hereafter;

g) Effluent from the production facility should be inactivated by validated chemical or physical means prior to discharge.

C. Category 3

At this level of physical containment the following objectives should be achieved:

a) Viable organisms should be handled in a production system which physically separates the process from the environment;

b) Exhaust gases should be treated to prevent the release of viable organisms;

c) Sample collection, addition of materials to a closed system and the transfer of viable organisms to another closed system should be done in a manner which prevents release;

d) Culture fluids should not be removed from the closed system unless the viable organisms have been inactivated by validated chemical or physical means;

e) Seals should be designed to prevent leakage or should be fully enclosed in ventilated housings;

f) Production systems should be located within a purpose built controlled area according to the requirements 6 *(a)* to *(k)* inclusive specified in the Table hereafter;

g) Effluent from the production facility should be inactivated by validated chemical or physical means prior to discharge.

EXAMPLES OF CONTAINMENT APPROACHES
FOR LARGE SCALE INDUSTRIAL APPLICATIONS OTHER THAN GILSP
(GOOD INDUSTRIAL LARGE SCALE PRACTICE)

Specifications	Containment Categories		
	1	2	3
1. Viable organisms should be handled in a system which physically separates the process from the environment (closed system)	Yes	Yes	Yes
2. Exhaust gases from the closed system should be treated so as to:	Minimise release	Prevent release	Prevent release
3. Sample collection, addition of materials to a closed system and transfer of viable organisms to another closed system, should be performed so as to:	Minimise release	Prevent release	Prevent release
4. Bulk culture fluids should not be removed from the closed system unless the viable organisms have been:	Inactived by validated means	Inactived by validated chemical or physical means	Inactived by validated chemical or physical means
5. Seals should be designed so as to:	Minimise release	Prevent release	Prevent release
6. Closed systems should be located within a controlled area	Optional	Optional	Yes, and purpose-built
a) Biohazard signs should be posted	Optional	Yes	Yes
b) Access should be restricted to nominated personnel only	Optional	Yes	Yes, via an airlock
c) Personnel should wear protective clothing	Yes work clothing	Yes	A complete change
d) Decontamination and washing facilities should be provided for personnel	Yes	Yes	Yes
e) Personnel should shower before leaving the controlled area	No	Optional	Yes
f) Effluent from sinks and showers should be collected and inactived before release	No	Optional	Yes
g) The controlled area should be adequately ventilated to minimise air contamination	Optional	Optional	Yes
h) The controlled area should be maintained at an air pressure negative to atmosphere	No	Optional	Yes
i) Input air and extract air to the controlled area should be HEPA filtered	No	Optional	Yes
j) The controlled area should be designed to contain spillage of the entire contents of the closed system	No	Optional	Yes
k) The controlled area should be sealable to permit fumigation	No	Optional	Yes
7. Effluent treatment before final discharge	Inactivated by validated means	Inactivated by validated chemical or physical means	Inactived by validated chemical or physical means

MANDATE OF THE *AD HOC* GROUP OF GOVERNMENT EXPERTS ON SAFETY AND REGULATIONS IN BIOTECHNOLOGY

The Committee for Scientific and Technological Policy has decided to set up a group of government experts on Safety and Regulations in Biotechnology with the following mandate:

1. The Group shall:

- Review country positions as to the safety in use of genetically engineered organisms at the industrial, agricultural and environmental levels, against the background of existing or planned legislation and regulations for the handling of micro-organisms;

2. In particular, the Group shall:

- Identify what *criteria* have been or may be adopted for the monitoring or authorization for production and use of genetically engineered organisms in:
  a) Industry;
  b) Agriculture;
  c) The environment;
- Explore possible ways and means for monitoring future production and use of genetically engineered organisms in:
  a) Industry;
  b) Agriculture;
  c) The environment.

3. The Group is requested to report to the Committee before June 1985.

4. This work should be a step towards better international harmonization of guidelines, codes of practice and/or regulations.

LIST OF PARTICIPANTS IN THE *AD HOC* GROUP OF GOVERNMENT EXPERTS ON SAFETY AND REGULATIONS IN BIOTECHNOLOGY

Dr. R. NOURISH   **CHAIRMAN**
H.M. Superintending Specialist Inspector
Technology and Air Pollution Division
Health and Safety Executive
(Bootle) Merseyside, United Kingdom

AUSTRALIA

Mr. P. FLAHERTY
Secretary, Recombinant DNA Monitoring Committee
Department of Industry, Technology & Commerce
Canberra

AUSTRIA

Prof. R. LAFFERTY
Institut für Biotechnologie
Mikrobiologie und Abfalltechnologie
Graz University
Graz

Mr. H. SCHWAB
Institut für Biotechnologie
Mikrobiologie und Abfalltechnologie
Graz University
Graz

BELGIUM

Professor H. COUSY
Katholieke Universiteit Leuven
Leuven

Prof. Dr. J. DE LEY
Microbiology Laboratory
Rijksuniversiteit
Ghent

Mrs. N. NOLARD
Institut d'Hygiène et d'Epidémiologie
Brussels

Mme. A.M. PRIEELS
Chargée de Mission
Services de Programmation de la Politique Scientifique
Brussels

CANADA

Dr. A. ALBAGLI
Senior Project Manager
National Research Council of Canada
Ottawa

Dr. J. FURESZ
Director, Bureau of Biologics
Department of Health and Welfare
Ottawa

Mr. T. McINTYRE
Advisor
Environmental Protection Service
Department of the Environment
Ottawa

Dr. D. B. SHINDLER
Manager-Biotechnology
Ministry of State for Science & Technology
Ottawa

DENMARK

Mrs. P. H. ANDERSEN   (Observer)
DVM, National Food Institute
Institute of Toxicology
Søborg

Prof. B. HARVALD (delegate until June 1984)
Chairman of the National Research Councils' Registration Committee
for Genetic Engineering
Copenhagen

Prof. E. LUND (delegate from June 1984 onwards)
Royal Vet. & Agricultural University of Copenhagen
Copenhagen

Mr. H. PEDERSEN   (Observer)
DVM, National Agency of Environmental Protection
Copenhagen

FINLAND

Prof. H. G. GYLLENBERG
Department of Microbiology
University of Helsinki
Helsinki

Dr. M. SARVAS
National Public Health Institute
Helsinki

FRANCE

Prof. G. BERNARDI
Directeur de Recherche
Laboratoire de Génétique Moléculaire
Université Paris VII
Paris

FRANCE *(cont'd)*

Mr. P. CAZALA
Ministère du Redéploiement Industriel et du Commerce Extérieur
Paris

Dr. N. LELONG
Chef de la Division Biotechnologie-Bioindustrie
Ministère du Redéploiement Industriel et du Commerce Extérieur
Paris

M. P. PRINTZ
Programme Mobilisateur Biotechnologies
Ministère de la Recherche et de la Technologie
Paris

GERMANY

Prof. Dr. W. FROMMER
Bayer AG
Wuppertal

Prof. Dr. M.A. KOCH
Bundesgesundheitsamt
(Federal Health Office)
Berlin

Dr. Peter LANGE
Bundesministerium für Forschung und Technologie,
(Ministry of Research & Technology)
Bonn

GREECE

Dr. G. TZOTZOS
Scientific Advisor
Ministry of Research and Technology
Department of Science Policy
Athens

IRELAND

Prof. S. DOONAN
Professor of Biochemistry
University College
Cork

ITALY

Prof. C. FRONTALI   (Observer)
Istituto Superiore di Sanità
Laboratorio di Biologia Cellulare
Rome

Prof. V. SGARAMELLA
Dipartimento di Genetica e Microbiologia
University of Pavia
Pavia

JAPAN

Mr. T. FUKUMIZU
Former Deputy Director
Bioindustry Office
Ministry of International Trade & Industry
Tokyo

Dr. H. HARADA
Professor
Institute of Biological Sciences
University of Tsukuba
Sakura-mura, Niihari-gun

Mr. R. HIGASHIUCHI
Deputy Director
Ministry of Health & Welfare
Tokyo

Mr. H. HIRAMATSU
Director
Bioindustry Office
Ministry of International Trade & Industry
Tokyo

Mr. N. INOUE
Counsellor
Ministry of International Trade & Industry
Tokyo

Mr. T. ITO
Deputy Director
Biology and Antibiotics Division
Ministry of Health & Welfare
Tokyo

Dr. T. TAKAHASHI, M.D.
Director, Life Sciences Division
Science and Technology Agency
Tokyo

Dr. K. TANAKA
Chairman of Development Promoting Committee
The Japan Association for Advanced Research of Pharmaceuticals
Tokyo

Mr. M. TANAKA
Former Director
Bioindustry Office
Ministry of International Trade & Industry
Tokyo

Dr. A. OYA
Director, Department of Virology & Rickettsiology
National Institute of Health
Tokyo

Dr. S. TSURU
Secretariat
Council of Agriculture, Forestry & Fisheries
Tokyo

JAPAN *(cont'd)*

Prof. H. UCHIDA
Advisor
University of Tokyo
Tokyo

Prof. I. WATANABE
Faculty of Hygiene
Kitazato University
Sagamihala-City

NETHERLANDS

Dr. A.W.J.J. BUIJS
Ministry of Housing,
Physical Planning and Environment
Leidschendam

Dr. Van EE
Gist Brocades N.V.
Delft

Prof. Dr. D.G. de HAAN
Laboratory for Microbiology
Rijksuniversiteit te Utrecht
Utrecht

Mr. M. C. KROON, M.C.
Ministry of Housing, Physical Planning
and Environment
Leidschendam

NORWAY

Dr. W. GUNDERSEN
Associate Professor
Institute of General Genetics
University of Oslo
Oslo

Mr. S. HAGEN
State Pollution Control Authority
Oslo

Mr. B. HAREIDE
Director, M.D.,
National Institute of Public Health
Oslo

Mr. K. S. NORBRAATHEN
State Pollution Control Authority
Oslo

PORTUGAL

Prof. L. ARCHER
Universidade Nova de Lisboa
Instituto Gulbenkian de Ciencia
Oeiras

SPAIN

Prof. A. ALBERT
Science Advisor
Direccion General de Política Científica
Madrid

Dr. R. REVILLA PEOREIRA
Ministerio de Industria y Energia
Madrid

SWEDEN

Dr. G. BRUNIUS
Swedish Recombinant DNA Advisory Committee
National Board of Occupational Safety & Health
Solna

SWITZERLAND

Dr. M. KÜENZI
Ciba-Geigy AG
Basel

TURKEY

Prof. M. BARA
Faculty of Sciences
Istanbul University
Istanbul

UNITED KINGDOM

Mr. B.P. AGER
Secretary, Advisory Committee on Genetic Manipulation
Health & Safety Executive
London

Mr. W.E.O. JONES
Health & Safety Executive
Medical Division
London

Mrs. M. PRATT
Plant Pathologist
Ministry of Agriculture, Fisheries and Food
Harpenden, Herts

Mr. J.F.A. THOMAS
Department of Environment
London

Mr. J. F. THORLEY
DISTA Products Ltd.
Liverpool

Mr. R. E. BENEDICK
Deputy Assistant Secretary
Environment, Health and Natural Resources
Department of State
Washington, D.C.

Mr. D. CLAY
Director, Office of Toxic Substances
Environmental Protection Agency
Washington, D.C.

Mr. J. COHRSSEN
Regulatory Counsel
Office of Science and Technology Policy
Executive Office of the President
Washington, D.C.

Dr. D.L. DULL
Office of Toxic Substances
Environmental Protection Agency
Washington, D.C.

Dr. J. R. FOWLE III
ORD Biotechnology Coordinator
Environmental Protection Agency
Washington, D.C.

Mr. I. FULLER
Director
Industrial Competitive Assessment
Office of the U.S. Trade Representative
Executive Office of the President
Washington, D.C.

Dr. W. J. GARTLAND, Jr.
Director
Office of Recombinant DNA Activities
National Institutes of Health
Bethesda, Maryland

Dr. E.L. KENDRICK
Acting Deputy Assistant Secretary
Department of Agriculture
Washington, D.C.

Dr. M. A. LEVIN
Office of Research and Development
Environmental Protection Agency
Washington, D.C.

Dr. C. MAZZA
Office of Toxic Substances
Environmental Protection Agency
Washington, D.C.

Dr. E. MILEWSKI
Office of Recombinant DNA Activities
National Institutes of Health
Bethesda, Maryland

USA *(cont'd)*

Dr. Henry I. MILLER, M.D.
Medical Officer
Food and Drug Administration
Rockville

Mr. M. L. SMITH
Executive Secretary
Biotechnology Group
Office of Policy Development
Executive Office of the President
Washington, D.C.

Mr. R. J. SMITH
Deputy Assistant Secretary
Bureau of Oceans and International Environmental and Scientific Affairs
Department of State
Washington, D.C.

Dr. S. TOLIN
Consulting Scientist
Department of Agriculture
Washington, D.C.

Mr. William J. WALSH, III
Co-ordinator for Biomedical Research & Health Affairs
Department of State
Washington, D.C.

Dr. F. E. YOUNG
Commissioner
Food & Drug Administration
Rockville

YUGOSLAVIA

Prof. Vladimir GLISIN
Institute for Biological Research
University of Belgrade
Belgrade

COMMISSION OF THE EUROPEAN COMMUNITIES

Dr. G. Del BINO
Environment, Consumer Protection & Nuclear Safety Directorate (DGXI)
Brussels

Mr. M. F. CANTLEY
Directorate-General for Science, Research & Development (DGXII)
CUBE (Concertation Unit for Biotechnology in Europe)
Brussels

Mr. T. GARVEY
Director
Internal Market & Industrial Affairs (DGIII)
Brussels

COMMISSION OF THE EUROPEAN COMMUNITIES *(cont'd)*

Mr. C. MANTEGAZZINI
Consultant to the Environment,
Consumer Protection & Nuclear Safety (DGXI)
Brussels

Dr. Ken SARGEANT
Directorate General for Science, Research & Development (DGXII)
CUBE (Concertation Unit for Biotechnology in Europe)
Brussels

Mr. F. SAUER
Internal Market & Industrial Affairs Directorate (DGIII)
Pharmaceutical Division
Brussels

Miss C. WHITEHEAD
Consultant to the Environment
Consumer Protection & Nuclear Safety Directorate (DGXI)
Brussels

CONSULTANT

Mr. W. ROSKAM
ELF Bio Recherche
Castanet Tolosan.
FRANCE

OECD SECRETARIAT

Miss Bruna Teso
Directorate for Science, Technology & Industry

GLOSSARY

Aerosol A suspension of fine liquid particles in a gas.

Allele Alternate forms of the same gene. For example, the genes responsible for eye colour (blue, brown, green, etc.) are alleles.

Amino Acid The building units of proteins; amino acids are linked together in a particular order which determines the character of different proteins.

Antigen A macromolecule (usually a protein or carbohydrate) which, when introduced into the body of a human or higher animal, stimulates the production of an antibody that reacts specifically with it.

Assay A technique that measures a biological response.

Auxotrophy Requirement by a mutant micro-organism for growth factors not needed by the corresponding wild type micro-organism.

Biological Containment Characteristics of an organism which limit its survival and/or multiplication in an environment.

Biomass All organic matter that grows by the photosynthetic conversion of solar energy.

Biopolymer Naturally occurring macromolecule that include proteins, nucleic acids, and polysaccharides.

Biosynthetic process The process by which chemical compounds are produced by a living organism either by synthesis or degradation.

Biota The flora and fauna of a region.

Cell Mass of living material surrounded by a membrane; the basic structural and functional unit of most organisms.

Cell Culture The *in vitro* growth of cells isolated from multicellular organisms. These cells are usually of one type.

Chloroplasts Cellular organelles where photosynthesis occurs.

Clone A collection of genetically identical cells or organisms which have been derived asexually from a common ancestor; all members of the clone have identical genetic composition.

Colonisation The establishment of a population within a new territory, e.g., the establishment of a novel colony of micro-organisms in the gastrointestinal tract.

Conjugation The one-way transfer of DNA between bacteria in cellular contact.

Cross-breeding To interbreed two varieties or breeds of the same species.

Culture fluid The medium in which the organism is grown.

Deamination Removal of amino (NH_2) groups.

Dehalogenation Removal of halogen (e.g. Cl_2, I_2) groups.

Denitration To remove nitric acid, nitrates, the nitro group, or nitrogen oxides.

DNA Deoxyribonucleic acid; polymer composed of deoxyribonucleotide units; genetic material of all organisms except RNA viruses.

Donor organism The organism from which DNA is taken to insert into the recipient or host organism in r-DNA constructions.

Ecosystem The complex of a community and its environment functioning as an ecological unit in nature.

Electron transfer proteins Proteins involved in the sequential transfer of electrons, (especially in cellular respiration) from an oxidisable substrate to molecular oxygen by a series of oxidation-reduction reactions.

Enzyme A protein that catalyses a chemical reaction.

Epidemiological Concerning the incidence, distribution and control of organisms, particularly a pathogen.

Eukaryotic The highly differentiated cell which is the unit of structure in animals, plants, protozoa, fungi and algae.

Fermentation An anaerobic bioprocess. Fermentation is used in various industrial processes for the manufacture of products such as alcohols, acids, and cheese by the action of yeasts, molds, and bacteria.

Flocculation The agglomeration of suspended material to form particles that will settle by gravity, as in the 'tertiary' treatment of waste materials.

Gene The basic unit of heredity; an ordered sequence of nucleotide bases, comprising a segment of DNA. A gene contains the sequence of DNA that encodes one polypeptide chain (via RNA).

Gene Probe A specific DNA or RNA sequence used to detect complementary sequences among nucleic acid molecules.

Genetic Material DNA, genes, chromosomes which constitute an organism's hereditary material; RNA in certain viruses.

Genome The genetic endowment of an organism or individual.

Host The organism into which donor DNA is inserted in r-DNA constructions; provides the major portion of the genome of the r-DNA organism; same as recipient.

In Vitro Literally, in glass; pertaining to a biological reaction taking place in an artificial apparatus; sometimes used to include the growth of cells from multicellular organisms under cell culture conditions. *In vitro* diagnostic products are products used to diagnose disease outside of the body after a sample has been taken from the body.

In Vivo Literally, in life; pertaining to a biological reaction taking place in a living cell or organism. *In vivo* products are products used within the body.

Insectary Place for the keeping or rearing of living insects.

Leaching The removal of a soluble compound such as an ore from a solid mixture by washing or percolating.

Lipopolysaccharide A water-soluble lipid-polysaccharide complex.

Metazoan Cell Cell from a multicellular (metazoan) organism rather than unicellular (protozoan) organism.

Micro-organism Microscopic living entity; micro-organisms can be viruses, procaryotes (e.g. bacteria) or eucaryotes (e.g. fungi). Also referred to as microbes.

Microcosm A community which is a representation of a larger system.

Microinjection The technique of introducing very small amounts of material (DNA or RNA molecules, enzymes, cytotoxic agents) into an intact cell through a microscopic needle penetrating the cell membrane.

Mitochondria Structures in higher cells that serve as the "powerhouse" for the cell, producing chemical energy.

Modulators of the immune system Non-antibody proteins released by primed lymphocytes on contact with antigen which act as intercellular mediators of the immune response.

Monoclonal antibodies Antibodies derived from a single source or clone of cells which recognize only one kind of antigen.

Mutagenesis The induction of mutation in the genetic material of an organism; researchers may use physical or chemical means to cause mutations that improve the production capabilities of organisms.

Mutation Any change that alters the sequence of bases along the DNA, changing the genetic material.

Non-viable Not capable of living, growing, or developing and functioning successfully.

Organism Any biological entity, cellular or non-cellular, with capacity for self-perpetuation and response to evolutionary forces; includes plants, animals, fungi, protists, prokaryotes, and viruses.

Pathogen A disease-producing agent, usually restricted to a living agent such as a bacterium or virus.

Pathogenic Capable of causing disease.

Phage A virus that multiplies in bacteria.

Phagocytosis The engulfing and (usually) destruction of particulate matter by cells (such as the leukocyte) that characteristically engulf foreign material and consume debris.

Phenotype The characteristics of an organism that results from the interaction of its genetic constitution with the environment.

Physical Containment Procedures or structures designed to reduce or prevent the release of viable organisms; degree of containment varies.

Pituitary A small oval endocrine organ that is attached to the infundibulum of the brain and produces various internal secretions directly or indirectly impinging on most basic body functions.

Plasmid An extrachromosomal, self-replicating, circular segment of DNA; plamids (and some viruses) are used as "vectors" for cloning DNA in bacterial "host" cells.

Polynucleotide A polymeric chain of compounds consisting of a ribose or deoxyribose sugar joined to a purine or pyrimidine base and a phosphate group and that are the basic structural units of DNA and RNA.

Polypeptide A long peptide, which consists of amino acids.

Population A body of individuals having a characteristic in common.

Prokaryotic The less differentiated cell which is the unit of structure in bacteria.

Protoplast A cell without a wall.

Reagent A substance that takes part in a chemical reaction.

Recipient Organism See "Host".

Retrovirus An animal virus with a glycoprotein envelope and an RNA genome that replicates through a DNA intermediate.

Ring cleavage The splitting of a compound in which the molecules are arranged cyclically (a closed chain) commonly consisting of five or six atoms although smaller and larger rings are known.

Scale-up The transition of a process from an experimental scale to an industrial scale.

Sclerotia Fungal survival structures capable of remaining dormant for long periods.

Screening To select organisms on the basis of a specific characteristic.

Secondary Metabolite Metabolite that is not required by the producing organism for its life-support system.

Selection A laboratory process by which cells or organisms are chosen for specific characteristics.

Single Cell Protein   Cells, or protein extracts, of micro-organisms grown in large quantities for use as human or animal protein supplements.

Spore   A dormant cellular form, derived from a bacterial or a fungal cell, that is devoid of metabolic activity and that can give rise to a vegetative cell upon germination; it is dehydrated and can survive for prolonged periods of time under drastic environmental conditions.

Storage Protein Genes   Genes coding for the major proteins found in plant seeds.

Substrate   A substance acted upon, for example, by an enzyme.

Symbiont   An organism living in symbiosis, usually the smaller member of a symbiotic pair of dissimilar size.

Symbiotic   Capable of living in an intimate association with a dissimilar organism in a mutually beneficial relationship.

Toxoid   Detoxified toxin, but with antigenic properties intact.

Transgenic animals   Animals into which DNA from another species are introduced by microinjection or retroviral infection.

Translocation   The exchange of parts between non-homologous chromosomes.

Vector   An agent of transmission; for example a DNA vector is a self-replicating molecule of DNA that transmits genetic information from one cell or organism to another. Plasmids (and some viruses) are used as "vectors" for DNA in bacterial cloning.

Viroid   Small pathogenic RNA molecule, apparently unable to code for protein, which depends on the host machinery for its replication.

Zygote   A cell formed by the union of two mature reproductive cells.

OECD SALES AGENTS
DÉPOSITAIRES DES PUBLICATIONS DE L'OCDE

ARGENTINA - ARGENTINE
Carlos Hirsch S.R.L.,
Florida 165, 4° Piso,
(Galeria Guemes) 1333 Buenos Aires
Tel. 33.1787.2391 y 30.7122

AUSTRALIA-AUSTRALIE
D.A. Book (Aust.) Pty. Ltd.
11-13 Station Street (P.O. Box 163)
Mitcham, Vic. 3132   Tel. (03) 873 4411

AUSTRIA - AUTRICHE
OECD Publications and Information Centre,
4 Simrockstrasse,
5300 Bonn (Germany)   Tel. (0228) 21.60.45
Local Agent:
Gerold & Co., Graben 31, Wien 1 Tel. 52.22.35

BELGIUM - BELGIQUE
Jean de Lannoy, Service Publications OCDE,
avenue du Roi 202
B-1060 Bruxelles   Tel. 02/538.51.69

CANADA
Renouf Publishing Company Limited/
Éditions Renouf Limitée Head Office/
Siège social – Store/Magasin :
61, rue Sparks Street,
Ottawa, Ontario KIP 5A6
Tel. (613)238-8985. 1-800-267-4164
Store/Magasin : 211, rue Yonge Street,
Toronto, Ontario M5B 1M4.
Tel. (416)363-3171
Regional Sales Office/
Bureau des Ventes régional :
7575 Trans-Canada Hwy., Suite 305,
Saint-Laurent, Quebec H4T 1V6
Tel. (514)335-9274

DENMARK - DANEMARK
Munksgaard Export and Subscription Service
35, Nørre Søgade, DK-1370 København K
Tel. +45.1.12.85.70

FINLAND - FINLANDE
Akateeminen Kirjakauppa,
Keskuskatu 1, 00100 Helsinki 10   Tel. 0.12141

FRANCE
OCDE/OECD
Mail Orders/Commandes par correspondance :
2, rue André-Pascal,
75775 Paris Cedex 16
Tel. (1) 45.24.82.00
Bookshop/Librairie : 33, rue Octave-Feuillet
75016 Paris
Tel. (1) 45.24.81.67 or/ou (1) 45.24.81.81
Principal correspondant :
Librairie de l'Université,
12a, rue Nazareth,
13602 Aix-en-Provence   Tel. 42.26.18.08

GERMANY - ALLEMAGNE
OECD Publications and Information Centre,
4 Simrockstrasse,
5300 Bonn   Tel. (0228) 21.60.45

GREECE - GRÈCE
Librairie Kauffmann,
28, rue du Stade, 105 64 Athens Tel. 322.21.60

HONG KONG
Government Information Services,
Publications (Sales) Office,
Beaconsfield House, 4/F.,
Queen's Road Central

ICELAND - ISLANDE
Snæbjörn Jónsson & Co., h.f.,
Hafnarstræti 4 & 9,
P.O.B. 1131 – Reykjavik
Tel. 13133/14281/11936

INDIA - INDE
Oxford Book and Stationery Co.,
Scindia House, New Delhi 1   Tel. 45896
17 Park St., Calcutta 700016   Tel. 240832

INDONESIA - INDONESIE
Pdin Lipi, P.O. Box 3065/JKT.Jakarta
Tel. 583467

IRELAND - IRLANDE
TDC Publishers – Library Suppliers
12 North Frederick Street, Dublin 1
Tel. 744835-749677

ITALY - ITALIE
Libreria Commissionaria Sansoni,
Via Lamarmora 45, 50121 Firenze
Tel. 579751/584468
Via Bartolini 29, 20155 Milano   Tel. 365083
Sub-depositari :
Ugo Tassi, Via A. Farnese 28,
00192 Roma   Tel. 310590
Editrice e Libreria Herder,
Piazza Montecitorio 120, 00186 Roma
Tel. 6794628
Agenzia Libraria Pegaso,
Via de Romita 5, 70121 Bari
Tel. 540.105/540.195
Agenzia Libraria Pegaso, Via S.Anna dei
Lombardi 16, 80134 Napoli.   Tel. 314180
Libreria Hœpli,
Via Hœpli 5, 20121 Milano   Tel. 865446
Libreria Scientifica
Dott. Lucio de Biasio "Aciou"
Via Meravigli 16, 20123 Milano   Tel. 807679
Libreria Zanichelli, Piazza Galvani 1/A,
40124 Bologna   Tel. 237389
Libreria Lattes,
Via Garibaldi 3, 10122 Torino   Tel. 519274
La diffusione delle edizioni OCSE è inoltre
assicurata dalle migliori librerie nelle città più
importanti.

JAPAN - JAPON
OECD Publications and Information Centre,
Landic Akasaka Bldg., 2-3-4 Akasaka,
Minato-ku, Tokyo 107   Tel. 586.2016

KOREA - CORÉE
Pan Korea Book Corporation
P.O.Box No. 101 Kwangwhamun, Seoul
Tel. 72.7369

LEBANON - LIBAN
Documenta Scientifica/Redico,
Edison Building, Bliss St.,
P.O.B. 5641, Beirut   Tel. 354429-344425

MALAYSIA - MALAISIE
University of Malaya Co-operative Bookshop
Ltd.,
P.O.Box 1127, Jalan Pantai Baru,
Kuala Lumpur   Tel. 577701/577072

NETHERLANDS - PAYS-BAS
Staatsuitgeverij
Chr. Plantijnstraat, 2 Postbus 20014
2500 EA S-Gravenhage   Tel. 070-789911
Voor bestellingen: Tel. 070-789880

NEW ZEALAND - NOUVELLE-ZÉLANDE
Government Printing Office Bookshops:
Auckland: Retail Bookshop, 25 Rutland Street,
Mail Orders, 85 Beach Road
Private Bag C.P.O.
Hamilton: Retail: Ward Street,
Mail Orders, P.O. Box 857
Wellington: Retail, Mulgrave Street, (Head
Office)
Cubacade World Trade Centre,
Mail Orders, Private Bag
Christchurch: Retail, 159 Hereford Street,
Mail Orders, Private Bag
Dunedin: Retail, Princes Street,
Mail Orders, P.O. Box 1104

NORWAY - NORVÈGE
Tanum-Karl Johan a.s
P.O. Box 1177 Sentrum, 0107 Oslo 1
Tel. (02) 801260

PAKISTAN
Mirza Book Agency
65 Shahrah Quaid-E-Azam, Lahore 3 Tel. 66839

PORTUGAL
Livraria Portugal,
Rua do Carmo 70-74, 1117 Lisboa Codex.
Tel. 360582/3

SINGAPORE - SINGAPOUR
Information Publications Pte Ltd
Pei-Fu Industrial Building,
24 New Industrial Road No. 02-06
Singapore 1953   Tel. 2831786, 2831798

SPAIN - ESPAGNE
Mundi-Prensa Libros, S.A.,
Castelló 37, Apartado 1223, Madrid-28001
Tel. 431.33.99
Libreria Bosch, Ronda Universidad 11,
Barcelona 7   Tel. 317.53.08/317.53.58

SWEDEN - SUÈDE
AB CE Fritzes Kungl. Hovbokhandel,
Box 16356, S 103 27 STH,
Regeringsgatan 12,
DS Stockholm   Tel. (08) 23.89.00
Subscription Agency/Abonnements:
Wennergren-Williams AB,
Box 30004, S104 25 Stockholm. Tel. 08/54.12.00

SWITZERLAND - SUISSE
OECD Publications and Information Centre,
4 Simrockstrasse,
5300 Bonn (Germany)   Tel. (0228) 21.60.45
Local Agent:
Librairie Payot,
6 rue Grenus, 1211 Genève 11
Tel. (022) 31.89.50

TAIWAN - FORMOSE
Good Faith Worldwide Int'l Co., Ltd.
9th floor, No. 118, Sec.2
Chung Hsiao E. Road
Taipei   Tel. 391.7396/391.7397

THAILAND - THAILANDE
Suksit Siam Co., Ltd.,
1715 Rama IV Rd.,
Samyam Bangkok 5   Tel. 2511630

TURKEY - TURQUIE
Kültur Yayinlari Is-Türk Ltd. Sti.
Atatürk Bulvari No: 191/Kat. 21
Kavaklidere/Ankara   Tel. 25.07.60
Dolmabahce Cad. No: 29
Besiktas/Istanbul   Tel. 160.71.88

UNITED KINGDOM - ROYAUME UNI
H.M. Stationery Office,
Postal orders only:
P.O.B. 276, London SW8 5DT
Telephone orders: (01) 622.3316, or
Personal callers:
49 High Holborn, London WC1V 6HB
Branches at: Belfast, Birmingham,
Bristol, Edinburgh, Manchester

UNITED STATES - ÉTATS-UNIS
OECD Publications and Information Centre,
Suite 1207, 1750 Pennsylvania Ave., N.W.,
Washington, D.C. 20006 - 4582
Tel. (202) 724.1857

VENEZUELA
Libreria del Este,
Avda F. Miranda 52, Aptdo. 60337,
Edificio Galipan, Caracas 106
Tel. 32.23.01/33.26.04/31.58.38

YUGOSLAVIA - YOUGOSLAVIE
Jugoslovenska Knjiga, Knez Mihajlova 2,
P.O.B. 36, Beograd   Tel. 621.992

Orders and inquiries from countries where Sales
Agents have not yet been appointed should be sent
to:
OECD, Publications Service, Sales and
Distribution Division, 2, rue André-Pascal, 75775
PARIS CEDEX 16.

Les commandes provenant de pays où l'OCDE n'a
pas encore désigné de dépositaire peuvent être
adressées à :
OCDE, Service des Publications. Division des
Ventes et Distribution. 2. rue André-Pascal. 75775
PARIS CEDEX 16.

69928-07-1986

OECD PUBLICATIONS, 2, rue André-Pascal, 75775 PARIS CEDEX 16 - No. 43691 1986
PRINTED IN FRANCE
(93 86 02 1) ISBN 92-64-12857-3